BrightRED Revision

Higher BIOLOGY

Cara Matthew and Clare Little

First published in 2009 by:

Bright Red Publishing Ltd
6 Stafford Street
Edinburgh
EH3 7AU

A CIP record for this book is available from the British Library

ISBN 978-1-906736-05-7

With thanks to Ken Vail Graphic Design, Cambridge (layout) and Anna Clark (copy-edit)

Cover design by Caleb Rutherford – eidetic

Illustrations by Ken Vail Graphic Design.

Acknowledgements

Every effort has been made to seek all copyright holders. If any have been overlooked then Bright Red Publishing will be delighted to make the necessary arrangements.

Bright Red Publishing would like to thank Emily Hooton and Graeme Morris for permission to reproduce the photographs on pages 32, 60, 61 and 87.

Printed and bound in Scotland by Scotprint.

CONTENTS

COURSE STRUCTURE

The Higher Biology course is divided into three units:

- Unit 1: Cell biology
- Unit 2: Genetics and adaptation
- Unit 3: Control and regulation

ASSESSMENT

The Higher Biology course is assessed in three ways:

- Each of the three units is assessed within your school using a National Assessment Bank test. NABs are set by the Scottish Qualifications Authority and consist of structured, short-answer questions at grade C. You must gain at least 26 marks out of a possible 40 (or 65%) to pass.

- Practical abilities are also assessed internally. You are required to write a report on one of the investigations that you have carried out.

- You will also take an externally-assessed written examination consisting of a paper lasting 2·5 hours. The examination has an allocation of 130 marks and is divided into three sections:

  - Section A is worth 30 marks and consists of 30 multiple-choice questions; knowledge and understanding is tested in about 20 of the questions and the remainder test problem-solving and practical abilities.

  - Section B is worth 80 marks; between 50 and 55 marks are designed to test knowledge and understanding, with the remainder of the marks allocated to problem-solving and practical abilities.

  - Section C contains two essay (or extended-response) questions, each marked out of ten. Within each question, you will have to choose between two essay titles. The first essay is a structured essay that is divided into parts, with the marks for each part being indicated. The second essay is open-ended and carries one mark for coherence and one mark for relevance.

EXAM HINTS

You do not need to answer the questions in order. Find a question that you can answer easily, so that you settle your nerves.

Timekeeping is important if you are to complete the whole paper. As a general rule, you should be taking about one minute per mark. So allowing ten minutes for settling at the start and checking your paper at the end, the timing for each section should be roughly:

- multiple-choice questions – 30 minutes
- structured questions – 1 hour 20 minutes
- each essay question – 15 minutes

The course award is graded A, B, C or D based on how well you do in the external examination. To gain a course award, you must also pass the three NABs (one for each unit) and complete the investigation report to the standard required by the SQA.

THE STRUCTURE AND AIM OF THIS BOOK

There is no short-cut to passing any course at Higher level. To obtain a good pass requires consistent, regular revision over the duration of the course. The aim of this revision book is to help you to achieve this success by providing you with a concise and engaging coverage of the Higher Biology course material. We recommend that you use this book in conjunction with your class notes, to revise each topic area, prepare for NABs and prelims, and in your preparation for the final exam.

The book is divided between the three units of the course. Within each section, there is a double-page spread on each of the sub-sections.

Each double-page spread:

- provides the key ideas and concepts of the sub-section in a logical and digestible manner.

- contains 'Internet links' or 'Don't forget' boxes that flag up vital pieces of knowledge that you need to remember and important things that you must be able to do.

- contains a 'Lets think about this' feature which will extend your knowledge and understanding of the subject, and provide additional interest. Sometimes there are questions to help you check your understanding.

REVISION TIPS

- Don't leave your revision until the last minute. Make up a revision schedule, giving yourself enough time to revise thoroughly, and stick to it. Be realistic – you should work around your other activities and remember that you do need to take time off to relax away from your books.

- Find somewhere to study that is quiet and uncluttered. You need space to spread out your work.

- Study for short periods (between 30 and 45 minutes) with short breaks in between to keep your level of concentration higher. Go out of the room where you are studying during each break as this will help you to be refreshed and ready for your next study session.

- Read over each sub-topic at a slower pace than you would usually do and ask yourself questions or read it out loud. Make sure that you understand what you have been reading – you only learn what you understand.

- It's often easier to remember facts if you talk about topics with a family member or a friend. So, find a study buddy who can ask you questions about your work.

- Practice makes perfect; do past-paper practice so that the exam format is as familiar as possible. There are only a few ways in which you can be asked the same question and you will see similar questions and diagrams appearing in many past papers. Doing a past paper against the clock will also help you to get your time management right.

- In the run up to the exams, eat plenty of fresh fruit and vegetables to keep your energy levels up, and make sure that you get a good night's sleep so that you are alert throughout the exam.

Good luck, and enjoy!

CELL STRUCTURE AND VARIETY

CELL ULTRASTRUCTURE

An electron microscope can be used to examine the structures (**organelles**) that are found inside cells. As each cell type has a different function, not all organelles are found in every cell type. You should be familiar with the main organelles shown in the diagram and table below.

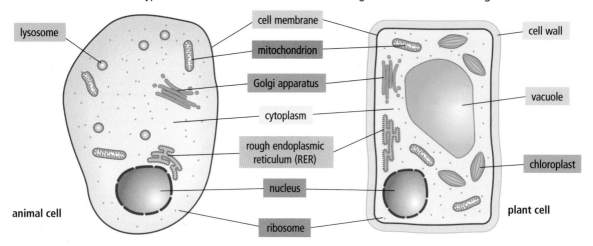

Cellular component (organelle)	Function	Location
nucleus	contains genetic material that controls the activities in the cell	plant and animal
rough endoplasmic reticulum (RER)	involved with transporting newly-made proteins in the cell	plant and animal
ribosome	site of protein synthesis	plant and animal
mitochondrion	aerobic respiration takes place here	plant and animal
Golgi apparatus	involved with processing and packaging newly-formed proteins before secretion from the cell	plant and animal
cell membrane	controls the entry and exit of substances	plant and animal
cytoplasm	site of chemical reactions	plant and animal
lysosome	contains digestive enzymes for breakdown of worn-out organelles and invading microbes	animal
cell wall	supports and protects the cell	plant
vacuole	contains cell sap – a mixture of salt, water and sugars	plant
chloroplast	site of photosynthesis	plant

UNICELLULAR ORGANISMS

Unicellular organisms are single cells. Therefore, each cell must contain all the organelles that the organism needs to survive. The diagrams below show two unicellular organisms and some of the organelles that enable them to function properly.

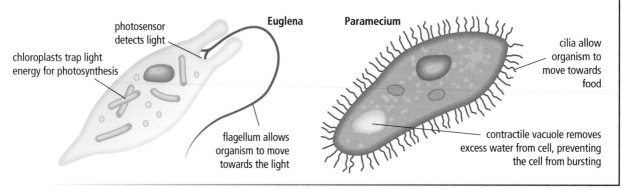

SPECIALISED CELLS IN A MULTICELLULAR ORGANISM

A **multicellular organism** is composed of many cells that work together to allow the organism to function. Different cell types contain only the organelles that they require to carry out their particular function within the organism. Some tissues are made up of the same type of cell, but there are tissues that contain different types of cell. For example, blood is composed of both red and white blood cells.

Type of cell		Specialised feature	Function
motor neurone		long cell process (axon) and many connections to other nerve cells	transmission of nerve impulses
goblet cell		cup-shaped cell with extensive Golgi apparatus	produce mucus that traps dirt and germs in respiratory tract
intestinal epithelial cell		finger-like extensions from cell surface (microvilli)	increase surface area for absorption
palisade mesophyll cell		many chloroplasts	main site of photosynthesis
guard cells		bean-shaped cells, arranged in pairs, with thickened cell wall on concave side	alter the size of the stoma to control gas exchange from a leaf
root hair cell		long cell extension to give a large surface area	absorption of water and mineral salts from the soil

DON'T FORGET

You should be familiar with the cell types you met in Standard Grade.

DON'T FORGET

Multicellular organisms carry out more complex functions than unicellular organisms.

 Look up http://cellsalive.com/cells/cell_model.htm

LET'S THINK ABOUT THIS

This cell is found in part of the kidney tubule.

(a) Complete the table below to identify the parts of the cell and their functions.

Part of cell	Name	Function
B		
	Golgi apparatus	
		transports proteins

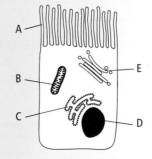

(b) This cell has many microvilli. Why?

LET'S THINK ABOUT THIS

You should be able to convert between millimetres (mm) and micrometres (µm).
Remember: 1 mm = 1000 µm and 1 µm = 0·001 mm
To change mm into µm, you multiply by 1000.
To change µm into mm, you divide by 1000.

1. What is 0·27 mm in µm?

2. Convert 57 µm into mm.

ABSORPTION AND SECRETION OF MATERIALS

FLUID MOSAIC MODEL

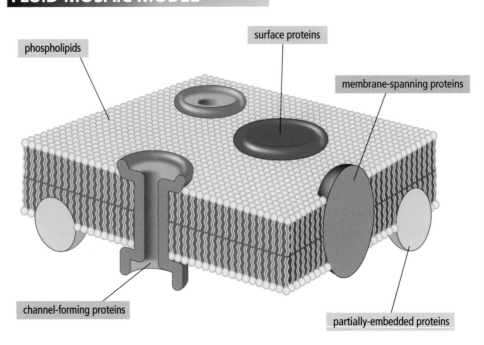

phospholipids

surface proteins

membrane-spanning proteins

channel-forming proteins

partially-embedded proteins

The plasma membrane is made up mainly of a double layer of constantly-moving (fluid) phospholipid molecules. **Proteins** of varying size are found scattered on and within the membrane, forming a mosaic pattern. Some of the functions carried out by these proteins are as follows:

● Channel-forming proteins create pores that allow small molecules to move passively through the membrane.

● Surface and partially-embedded proteins can act as receptor sites for antibodies and hormones, or act as enzymes.

● Some proteins that span the phospholipid bilayer act as carrier proteins for active transport.

The plasma membrane is said to be **selectively permeable** as small molecules, such as oxygen, carbon dioxide and water, pass through freely, while larger molecules (such as glucose, amino acids and urea) move through more slowly. Even larger molecules, for example proteins, are unable to pass through.

DIFFUSION

DON'T FORGET

Diffusion and osmosis are examples of **passive transport** and do not use energy.

Diffusion is the movement of molecules or ions from a region of high concentration to a region of low concentration. Diffusion is one process that allows substances to enter a cell and waste products to leave.

 Look up http://www.educypedia.be/education/biologycelldiffusion.htm

OSMOSIS

Osmosis is the movement of **water molecules** from a region of high water concentration to a region of low water concentration through a selectively-permeable membrane.

- When two solutions have different water concentrations, the one with the lower water concentration is said to be **hypertonic**. Marine organisms are surrounded by a hypertonic solution (sea water).

- Where the water concentration of two solutions is equal, the solutions are **isotonic**.

- A **hypotonic solution** has a higher water concentration than any it is being compared to. Animals living in rivers are surrounded by a hypotonic solution (river water).

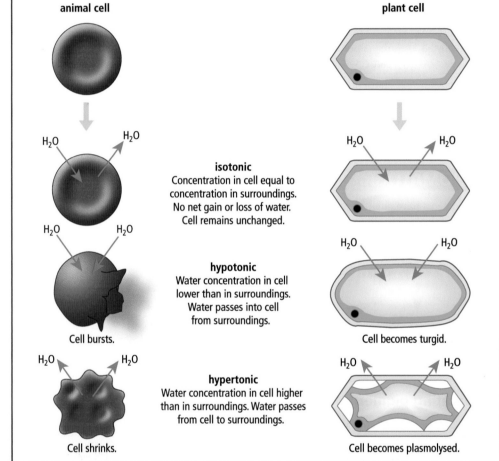

animal cell

H_2O H_2O

isotonic
Concentration in cell equal to concentration in surroundings. No net gain or loss of water. Cell remains unchanged.

H_2O H_2O

hypotonic
Water concentration in cell lower than in surroundings. Water passes into cell from surroundings.

Cell bursts.

H_2O H_2O

hypertonic
Water concentration in cell higher than in surroundings. Water passes from cell to surroundings.

Cell shrinks.

plant cell

H_2O H_2O

H_2O H_2O

Cell becomes turgid.

H_2O H_2O

Cell becomes plasmolysed.

DON'T FORGET

- Hypertonic – lower water concentration
- Hypotonic – higher water concentration

 Look up http://www.bcscience.com/bc8/pgs/quiz_section1.3.htm

ACTIVE TRANSPORT

Active transport is the movement of ions and molecules across the cell membrane **against a concentration gradient**. Molecules are moved across the membrane from a low to a high concentration by carrier proteins. As energy is required for this process, **ATP** (produced in aerobic respiration) must be available. Therefore, any factor that affects a cell's ability to produce ATP also affects the rate of active transport. These factors include: glucose concentration, oxygen concentration and temperature.

LET'S THINK ABOUT THIS

Refer to your investigation notes on osmosis in potato tissue or visking tubing as a practical example of the process. You may be asked to explain each stage in a data-handling question.

PHOTOSYNTHESIS – LEAF PIGMENTS AND SPECTRA

LEAF PIGMENTS

For light energy to be used to drive photosynthesis, it must be absorbed. Chemicals that absorb visible light are called **pigments**. The pigments involved in photosynthesis are found inside the **grana** of chloroplasts.

Pigment	Function
chlorophyll a	This absorbs red and blue light, and is the only pigment that can participate directly in light reactions.
chlorophyll b	These are accessory pigments which absorb other wavelengths of light. Captured energy is passed on to chlorophyll a. The accessory pigments are important as they increase the energy available for photosynthesis.
xanthophyll	
carotene	

DON'T FORGET

In green leaves, green light is reflected from the leaf.

CHROMATOGRAPHY

Leaf pigments have different levels of solubility and can be separated using **chromatography**; the most soluble pigment travels the greatest distance.

Extracting pigments

Grind the leaves in acetone ⸺ Bursts cells to release the pigments dissolved in the acetone

↓

Filter acetone mixture ⸺ Removes any insoluble cell debris

Separating pigments

Apply spots of the pigment extract to chromatography paper.

↓

Place the chromatography paper in a **solvent** (petroleum ether) with the pigment spots above the solvent surface. Leave to allow time for solvent to run up the paper and pigments to separate.

↓

Remove paper and record both the colour of the pigment and the distance each pigment has travelled from the start point.

Look up
www.lpscience.fatcow.com
/jwanamaker/animations
/Chrom%26Elpho.html

Look up
www.phschool.com/science/
biology_place/labbench/
lab4/pigsep.html

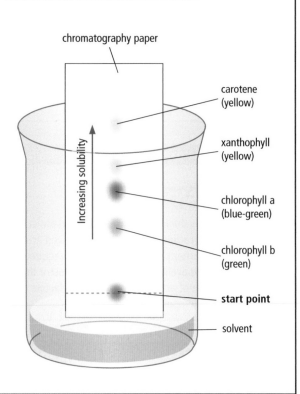

chromatography paper

carotene (yellow)

xanthophyll (yellow)

chlorophyll a (blue-green)

chlorophyll b (green)

start point

solvent

Increasing solubility

ABSORPTION AND ACTION SPECTRA

Absorption spectra for leaf pigments

The ability of the leaf pigments to absorb different wavelengths (colours) of light can be measured using a spectrophotometer. The graph produced is called an **absorption spectrum**.

The peaks on the graphs below show where absorption of light is greatest.

Chlorophyll a

Absorption of light by chlorophyll a is greatest in red and blue wavelengths.

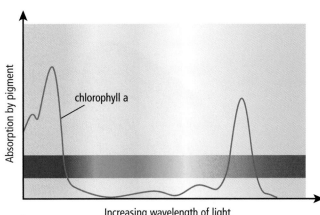

Chlorophyll b and carotene

Peak absorption by the accessory pigments is at different wavelengths to chlorophyll a. This extends the range of wavelengths that can be absorbed, so more light energy can be captured for photosynthesis.

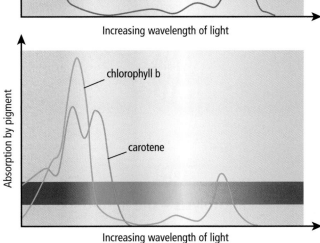

Action spectra for leaf pigments

An action spectrum is a graph showing the effect of different wavelengths of light on the rate of photosynthesis.

The peaks here show that the rate of photosynthesis is greatest within the red and blue wavelengths (absorbed by chlorophyll a) but remains high in the wavelengths absorbed by the other leaf pigments.

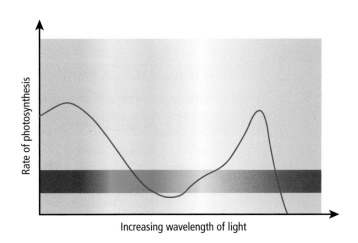

LET'S THINK ABOUT THIS

As well as being **absorbed**, light may be **reflected** or **transmitted**. Using the action spectrum, you should be able to explain why accessory pigments must be used during photosynthesis.

CHLOROPLAST STRUCTURE AND PHOTOSYNTHESIS

STRUCTURE OF A CHLOROPLAST

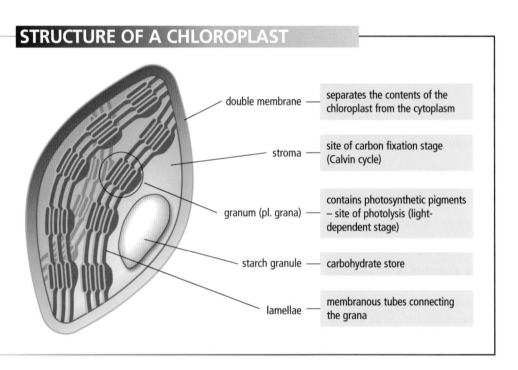

double membrane — separates the contents of the chloroplast from the cytoplasm

stroma — site of carbon fixation stage (Calvin cycle)

granum (pl. grana) — contains photosynthetic pigments – site of photolysis (light-dependent stage)

starch granule — carbohydrate store

lamellae — membranous tubes connecting the grana

STAGE ONE

Photolysis – the light-dependent stage

Photolysis takes place inside the **grana** of the chloroplasts. It is a light-dependent (photochemical) reaction.

Light energy is used in two ways:

1. It is trapped by the leaf pigments and used to split **water molecules** into **hydrogen** and **oxygen**.

2. It is used to convert **ADP + P$_i$** into **ATP**.

Hydrogen that is produced binds to an acceptor molecule called **NADP** to form **NADPH$_2$**. NADPH$_2$ leaves the grana to enter the stroma of the chloroplast for the second stage of photosynthesis, carbon fixation. The oxygen produced diffuses out of the cell as a waste product.

DON'T FORGET

Both hydrogen and ATP will be used in the next stage of photosynthesis: carbon fixation.

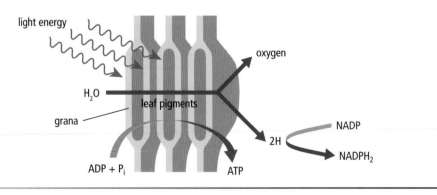

light energy

oxygen

H$_2$O

leaf pigments

grana

2H — NADP

NADPH$_2$

ADP + P$_i$

ATP

 Look up http://highered.mcgraw-hill.com/sites/0072437316/student_view0/ chapter10/animations.html

STAGE TWO

Carbon fixation – Calvin cycle

The second stage of photosynthesis takes place in the **stroma** of the chloroplast. It is a temperature-dependent (thermochemical) pathway that involves a cycle of several enzyme-controlled reactions.

Carbon dioxide (1C), gained by gas exchange from the air, diffuses into the stroma and enters **carbon fixation** by binding with molecules of **ribulose biphosphate (RuBP)** (5C).

As the cycle progresses, ATP and $NADPH_2$, both produced during photolysis, are required to allow the conversion of **glycerate phosphate (GP)** (3C) into **triose phosphate** (3C). Two triose phosphate molecules leave the cycle to produce glucose (6C). Most triose phosphate molecules remain in the cycle, where they are used to regenerate RuBP.

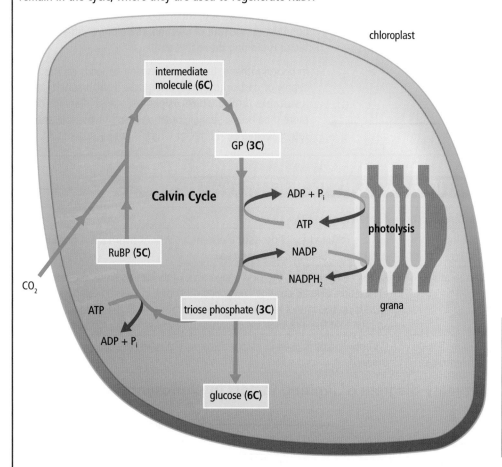

DON'T FORGET

You must be able to name the carbon dioxide acceptor molecule in carbon fixation and identify where hydrogen enters carbon fixation.

LET'S THINK ABOUT THIS

About 50% of the sugar produced in photosynthesis is used by the plant as a source of energy.

The other 50% is used to build proteins, fats and carbohydrates. Plants usually make more sugar than they require for respiration and production of cellular components. What does the plant do with this excess?

 Look up http://highered.mcgraw-hill.com/sites/0070960526/student_view0/chapter5/animation_quiz_1.html

 Look up http://faculty.nl.edu/jste/calvin_cycle.htm

LIMITING FACTORS IN PHOTOSYNTHESIS

LIMITING FACTORS

Any factor that can prevent a reaction progressing at its fastest rate is called a **limiting factor**. The graph below shows the effect of limiting factors.

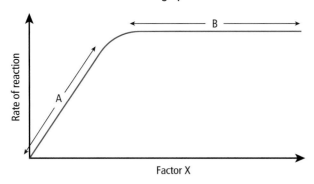

In part A of the graph, factor X is limiting the rate of the reaction (when X is increased, the rate of reaction increases). In part B of the graph, increasing factor X has no effect on the rate of reaction. Therefore, a different factor is limiting the reaction rate.

In photosynthesis, the rate of reaction depends on the supply of **carbon dioxide**, **adequate light intensity** and a **suitable temperature**. As photosynthesis depends on more than one essential condition being favourable, the rate is limited by the factor which is in shortest supply.

LIMITING FACTORS IN PHOTOSYNTHESIS

Limiting factor	Explanation
light intensity	Light is required for photolysis. If absent, no ATP or hydrogen is produced and carbon fixation cannot proceed.
temperature	At low temperatures, the slow movement of molecules limits the rate of reaction. Like all enzyme-controlled reactions, increasing temperature speeds up the rate of reaction in photosynthesis, until the temperature is so high that the enzymes are denatured.
carbon dioxide concentration	Carbon dioxide is used in carbon fixation. If absent, the cycle cannot proceed.

Summary diagram of limiting factors in photosynthesis

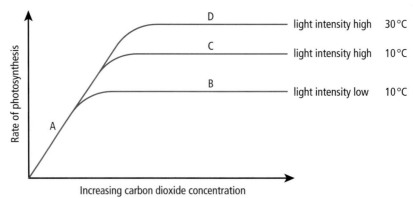

D	light intensity high 30 °C
C	light intensity high 10 °C
B	light intensity low 10 °C

> **DON'T FORGET**
>
> When interpreting graphs like this, make sure you are comparing two lines that show an alteration in just one variable. Lines C and D can be compared as they vary only in temperature. Lines B and D cannot be compared as they vary in both light intensity and temperature.

In part A of the graph, CO_2 concentration must be the limiting factor, as increasing the CO_2 concentration causes the rate of photosynthesis to rise. However, where the graph flattens out, increasing the concentration of CO_2 has no effect, and another limiting factor must be controlling the reaction rate. Comparison of curves B and C shows that light intensity is a limiting factor, as the rate of reaction is increased when the light intensity is increased from low to high. Similarly, comparison of curves C and D shows that temperature is a limiting factor. Here, increasing the temperature from 10°C to 30°C causes the reaction rate to increase.

SUMMARY OF PHOTOSYNTHESIS

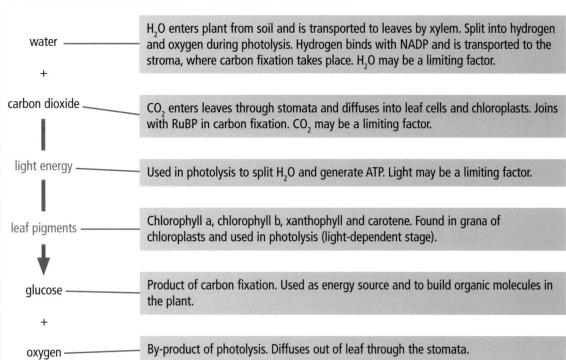

water — H_2O enters plant from soil and is transported to leaves by xylem. Split into hydrogen and oxygen during photolysis. Hydrogen binds with NADP and is transported to the stroma, where carbon fixation takes place. H_2O may be a limiting factor.

+

carbon dioxide — CO_2 enters leaves through stomata and diffuses into leaf cells and chloroplasts. Joins with RuBP in carbon fixation. CO_2 may be a limiting factor.

light energy — Used in photolysis to split H_2O and generate ATP. Light may be a limiting factor.

leaf pigments — Chlorophyll a, chlorophyll b, xanthophyll and carotene. Found in grana of chloroplasts and used in photolysis (light-dependent stage).

glucose — Product of carbon fixation. Used as energy source and to build organic molecules in the plant.

+

oxygen — By-product of photolysis. Diffuses out of leaf through the stomata.

⚙ LET'S THINK ABOUT THIS

You should be able to:

1. Identify the number of carbon atoms in each of the following molecules: carbon dioxide, RuBP, GP, glucose.

2. Explain why ATP and hydrogen must be transferred from photolysis to carbon fixation.

3. State the exact location of (a) photolysis and (b) carbon fixation.

⚙ LET'S THINK ABOUT THIS

The graphs below show the effect of temperature on photosynthesis. Can you identify which graph shows photolysis and which one is carbon fixation?

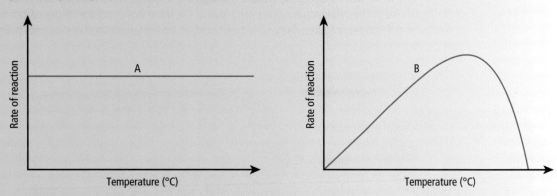

As photolysis does not require the presence of enzymes, increasing temperature has no effect on the rate of reaction (graph A). Carbon fixation is an enzyme-driven pathway, and so increasing the temperature increases the rate of reaction until the point where enzymes are denatured and the reaction rate drops to zero (graph B).

15

ENERGY RELEASE AND THE ROLE OF ATP

RESPIRATORY SUBSTRATES AND THE USES OF ENERGY

Molecules that can be broken down to release energy in respiration are called **respiratory substrates**. Glucose is the main respiratory substrate, although fats and proteins can be used if glucose is absent. The energy released is used to fuel cellular processes such as protein synthesis, contraction of muscle, active transport, DNA replication and carbon fixation.

ATP AND ADP

The series of reactions that make up respiration result in chemical energy being transferred to a molecule called **ATP**, adenosine triphosphate. ATP is a source of energy that can be used immediately by cells. During respiration, ATP is made when a bond forms between an **inorganic phosphate** (P_i) and **ADP**, adenosine diphosphate. The energy is stored in the bond. When the bond is subsequently broken, the energy is released and used in cellular processes.

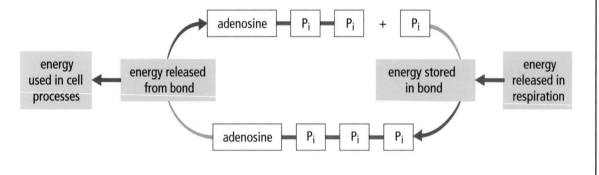

MITOCHONDRIA

Mitochondria are known as the powerhouses of the cell, because they are the main site of ATP synthesis. A mitochondrion consists of two membranes (a smooth outer membrane and a folded inner membrane) surrounding a central **matrix**. The Krebs cycle takes place in the matrix (see page 19). Most ATP is generated by the cytochrome system, which is present on the folds of the inner membrane (**cristae**). Folding increases the membrane's surface area, meaning that more of the molecules of the cytochrome system are present. The more folds, and the longer each fold is, the faster ATP can be produced. Very active cells contain lots of large mitochondria with many long cristae.

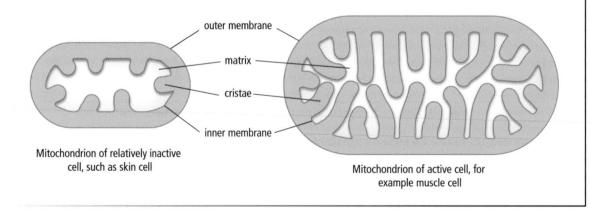

Mitochondrion of relatively inactive cell, such as skin cell

Mitochondrion of active cell, for example muscle cell

YEAST DEHYDROGENASE INVESTIGATION

Yeast cells contain the enzyme **dehydrogenase**, which removes hydrogen from molecules during respiration. Loss of hydrogen can be detected using **resazurin dye**, which changes colour from blue to pink and then to colourless as hydrogen is lost from the respiration pathway and added to the dye.

The table below shows the results of an experiment where resazurin dye, yeast and glucose solution were mixed together and left to react in a warm waterbath.

Test tube	Contents of test tube	Colour change	Explanation
A	resazurin dye boiled yeast suspension glucose solution	stays blue	Boiling the yeast suspension denatures the enzyme dehydrogenase, and no reaction occurs.
B	resazurin dye live yeast suspension glucose solution	blue → colourless	Dehydrogenase in the yeast cells catalyses oxidation of glucose. The hydrogen released during respiration is added to the resazurin dye.
C	resazurin dye water live yeast suspension	blue → pink	Although no glucose is present to act as respiratory substrate, the yeast cells do contain a small quantity of food which acts as a respiratory substrate. Therefore, some oxidation occurs and the dye becomes partially reduced.

LET'S THINK ABOUT THIS

Thinking about the yeast dehydrogenase experiment:

1. Why should all solutions be pre-incubated in a warm waterbath before being added together?

2. How could you modify the experiment to ensure that no stored carbohydrate remained in the yeast cells before using them?

LET'S THINK ABOUT THIS

Despite the fact that ATP is being made all the time in cells, the total mass in an organism is relatively constant. It is made when required and used almost immediately. To show that ATP – and not glucose – can be used as a direct source of energy by the cell, glucose and ATP solutions can be dripped onto teased-out muscle fibres. You should be able to describe what happens in each case.

Muscle cells do not contract when glucose solution is dripped onto them, but if ATP solution is used they contract immediately.

RESPIRATION

Respiration is a series of enzyme-controlled reactions.

STAGE ONE – GLYCOLYSIS

Glycolysis takes place in the **cytoplasm** of every living cell. No oxygen is required.

The conversion of 2ATP → 2ADP + 2P_i provides enough energy to begin the breakdown of glucose molecules. One (6C) **glucose** molecule is broken down into two (3C) **pyruvic acid** molecules. As four ATP molecules are produced during this reaction, there is a net gain of two ATP (4ATP produced – 2ATP used). Hydrogen is also released and picked up by a hydrogen carrier called **NAD** to make **NADH**.

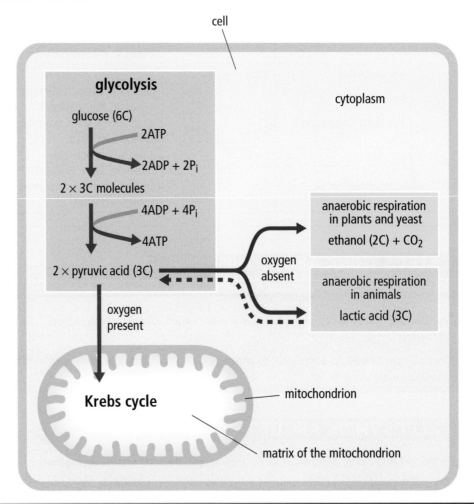

DON'T FORGET

Anaerobic respiration in animals is a reversible reaction. Lactic acid is converted back to pyruvic acid if the oxygen debt is repaid.

DON'T FORGET

Glycolysis results in a net gain of two ATPs.

 Look up www.mc.maricopa.edu/~johnson/animations/bread_glycolysis.swf

ANAEROBIC RESPIRATION

In the absence of oxygen, anaerobic respiration takes place. Anaerobic respiration produces only the **two ATPs** released in glycolysis. In plants and yeast, pyruvic acid is converted to **ethanol and carbon dioxide**. In animals, **lactic acid** is produced.

STAGE TWO – THE KREBS CYCLE

The Krebs cycle takes place in the **matrix of mitochondria**. It can only proceed if oxygen is present, as oxygen is the carrier molecule which combines with waste carbon atoms to form carbon dioxide.

On entering a mitochondrion, pyruvic acid is broken down to produce **acetyl co-enzyme A**, a 2C molecule. This enters the cycle by combining with a 4C molecule to produce **citric acid** (6C). As the cycle proceeds, carbon atoms are released and picked up by oxygen to form carbon dioxide, and hydrogen ions are released and picked up by **NAD** to form **NADH**. NADH carries the hydrogen ions to the third stage of respiration, the cytochrome system, which takes place on the inner membrane of the mitochondrion.

DON'T FORGET

Every time a carbon atom is lost, oxygen combines with it to form CO_2.

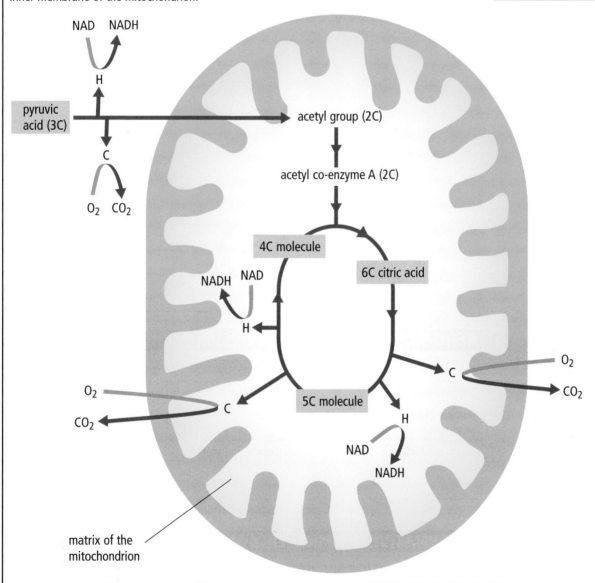

matrix of the mitochondrion

 Look up http://scholar.hw.ac.uk/site/biology/activity3.asp

LET'S THINK ABOUT THIS

For each glucose molecule, the Krebs cycle will go round twice – once for each of the two pyruvic acid molecules at the end of glycolysis.

You must be able to identify the named molecules in the respiratory pathway and state the number of carbon atoms that each contains.

RESPIRATION – THE CYTOCHROME SYSTEM AND ALTERNATIVE RESPIRATORY SUBSTRATES

THE CYTOCHROME SYSTEM

The cytochrome system is found on the **cristae of the mitochondria** and is the site of **oxidative phosphorylation**. Hydrogen is transferred to the cytochrome system by NADH. It is then passed along a series of hydrogen carriers. This is an **aerobic process**, as the final hydrogen acceptor is oxygen, which binds with hydrogen to produce water as a waste product. Every time a pair of hydrogen ions passes through the cytochrome system, three ATPs are produced. A total of **36 ATPs** per glucose molecule are made in the cytochrome system (plus two from glycolysis).

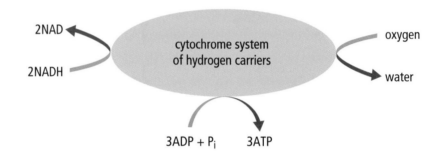

2NAD

2NADH

cytochrome system of hydrogen carriers

oxygen

water

$3ADP + P_i$ $3ATP$

AEROBIC AND ANAEROBIC RESPIRATION COMPARED

	Aerobic respiration	Anaerobic respiration
oxygen required	yes	no
number of ATP molecules produced	38	2
waste products	carbon dioxide and water	in plants and yeast, ethanol and carbon dioxide
		in animals, lactic acid

ALTERNATIVE RESPIRATORY SUBSTRATES

When glucose supplies have been exhausted, fats and proteins are used as respiratory substrates.

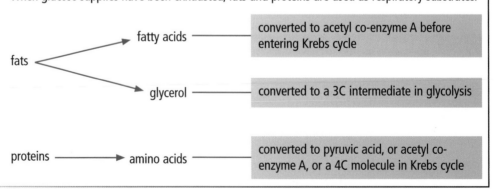

fats → fatty acids — converted to acetyl co-enzyme A before entering Krebs cycle

fats → glycerol — converted to a 3C intermediate in glycolysis

proteins → amino acids — converted to pyruvic acid, or acetyl co-enzyme A, or a 4C molecule in Krebs cycle

SUMMARY EQUATION FOR AEROBIC RESPIRATION

glucose ———————— in plants, product of photosynthesis; in animals, comes from the diet (main respiratory substrate)

+

oxygen ———————— enters organism by gas exchange with the air; used to remove both carbon atoms in the Krebs cycle and hydrogen from the cytochrome system

↓

carbon dioxide ———————— produced during the Krebs cycle when carbon atoms are lost and bind with oxygen

+

water ———————— produced at the end of the cytochrome system when hydrogen binds with oxygen

+

energy ———————— a total of 38 ATPs are produced per glucose molecule: two ATP during glycolysis and 36 ATPs from the cytochrome system

⚙ LET'S THINK ABOUT THIS

Aerobic respiration is the more efficient pathway. Compare 38 ATPs produced in aerobic respiration with only two ATPs in anaerobic respiration. What happens to the rest of the energy in anaerobic respiration? It remains as chemical energy stored in either ethanol (in plants and yeast) or lactic acid (in animals).

⚙ LET'S THINK ABOUT THIS

You should be able to demonstrate detailed knowledge of the reactions involved in aerobic and anaerobic respiration.

1. Complete the following table by placing a tick or a cross in each box, depending on whether the statement is correct.

Statement	Stage of aerobic respiration			anaerobic respiration
	glycolysis	Krebs cycle	oxidative phosphorylation	
oxygen required				
carbon dioxide produced				
hydrogen ions attach to NAD				
occurs in the matrix of the mitochondria				
occurs in the cytoplasm				
occurs on the cristae of the mitochondria				
produces ATP				
ethanol or lactic acid produced				

2. Explain the meaning of the term 'oxygen debt'.

3. What is the final hydrogen carrier molecule in aerobic respiration?

4. How many carbon atoms are present in a molecule of each of the following: glucose, pyruvic acid, acetyl co-enzyme A and citric acid?

STRUCTURE AND VARIETY OF PROTEINS

STRUCTURE OF PROTEINS

All proteins contain the elements **carbon**, **hydrogen**, **oxygen** and **nitrogen**. These elements combine to form an **amino acid**, the basic unit of a protein. There are approximately 20 naturally-occurring amino acids, giving rise to a huge variety of proteins.

Primary structure

Amino acids are brought together in a pre-determined order to form a chain called a **polypeptide**. In the chain, amino acids are held together by **peptide bonds**.

amino acid

peptide bond

Secondary structure

Once the primary structure has formed, the polypeptide becomes coiled and is held together by other bonds.

amino acid   bond

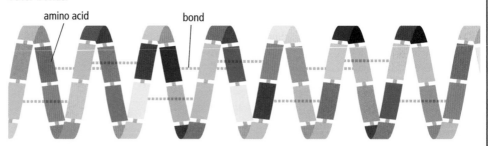

Tertiary structure

Finally, the polypeptide forms sheets (fibrous proteins) or is wound up to form a ball (globular proteins). Each protein's shape is maintained by bonds.

polypeptide   polypeptide

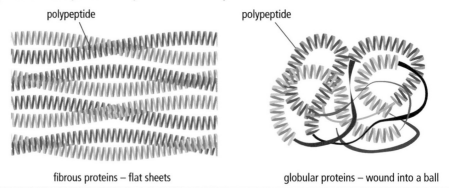

fibrous proteins – flat sheets   globular proteins – wound into a ball

VARIETY OF PROTEINS

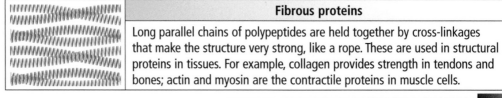

	Fibrous proteins
	Long parallel chains of polypeptides are held together by cross-linkages that make the structure very strong, like a rope. These are used in structural proteins in tissues. For example, collagen provides strength in tendons and bones; actin and myosin are the contractile proteins in muscle cells.

contd

VARIETY OF PROTEINS contd

Globular proteins		
Polypeptide chains are wound into a ball (like a tangled ball of wool), with the shape maintained by weak bonds.		
Type of globular protein	**Example**	**Function**
(a) structural	cell membrane proteins	Membrane proteins are involved in several important processes in the membrane, including support.
(b) enzymes	amylase	Shape provides an active site into which a substrate fits.
(c) hormones	insulin	Chemical messengers that target specific tissues to exert an effect.
(d) antibodies		Y-shaped proteins that provide two binding sites for antigens.
(e) conjugated proteins	haemoglobin	These are globular proteins bound to non-protein chemicals. Conjugated proteins have a range of functions in cells, including transport of molecules. Haemoglobin binds to oxygen in the blood.

Look up http://biology.about.com/od/molecularbiology/a/aa101904a.htm

LET'S THINK ABOUT THIS

The diagram shows one type of protein.

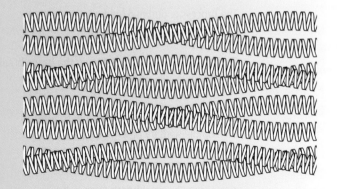

(a) Identify the type of protein.

(b) Give one example of this type of protein and state one way in which it is suited to its function.

(c) Haemoglobin is a conjugated protein. What does the term 'conjugated' mean?

(d) Explain why the shape of an enzyme molecule is of importance.

LET'S THINK ABOUT THIS

You should be able to describe the role of globular proteins in the cell membrane. Refer back to page 8 to review membrane structure and function.

NUCLEIC ACIDS

DNA

Deoxyribonucleic acid (DNA) is a double-stranded molecule made up of subunits called nucleotides.

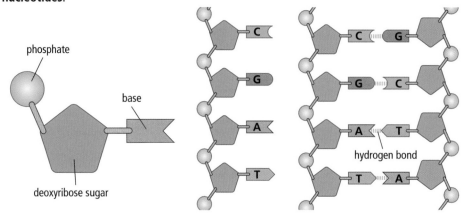

phosphate

base

deoxyribose sugar

hydrogen bond

Each nucleotide consists of a **phosphate** group, **deoxyribose** sugar and a **base**. The four different types of nucleotide each have a different base: **adenine**, **thymine**, **guanine** and **cytosine**. A chemical bond forms between the phosphate group of one nucleotide and the deoxyribose sugar of the next nucleotide, producing a strong strand.

Complementary base pairing occurs between the bases of two DNA strands. The bases are held together by hydrogen bonds, giving a double-stranded DNA molecule which then twists to form a double helix. In DNA, adenine always bonds with thymine (**A–T**), and cytosine always bonds with guanine (**C–G**).

DNA and replication

So that daughter cells receive the correct genetic information for normal metabolism, DNA must be copied before either **mitosis** or **meiosis** begins. This duplication of DNA is known as **replication**.

In order for replication to occur, the nucleus must contain:

- DNA
- the four types of nucleotide
- ATP
- enzymes

The process is outlined below.

1. The DNA molecule unwinds.

2. The hydrogen bonds between the bases break and the molecule 'unzips', exposing the bases of both DNA strands.

3. Free DNA nucleotides move in to form complementary base pairs with the exposed bases, and hydrogen bonds form between them.

4. Strong chemical bonds form between the phosphate and deoxyribose sugar of adjacent nucleotides.

5. The new daughter DNA molecules wind up, each forming a double helix.

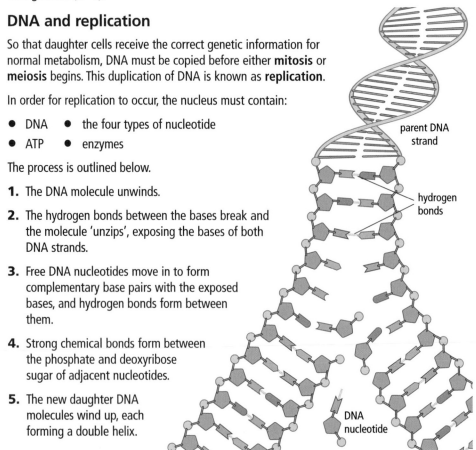

parent DNA strand

hydrogen bonds

DNA nucleotide

DNA contd

Replication is said to be **semi-conservative**, as the two daughter molecules each contain an original parent strand and a newly-formed strand.

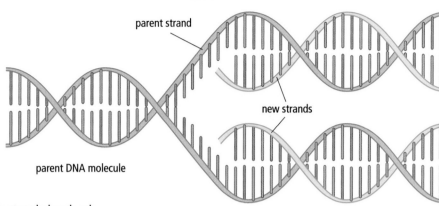

parent strand

new strands

parent DNA molecule

two daughter molecules

RNA

Ribonucleic acid (RNA) is a single-stranded molecule made of nucleotides.

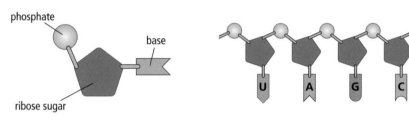

phosphate

base

ribose sugar

U   A   G   C

Each nucleotide consists of a phosphate group, ribose sugar and a base. There are four different bases: adenine, **uracil**, guanine and cytosine. Strong chemical bonds form between the phosphate group of one nucleotide and the ribose of the next nucleotide to form an RNA molecule.

You should be familiar with the following types of RNA, both of which are involved in protein synthesis (covered on pages 26–29):

- Messenger RNA (mRNA) is formed during transcription of DNA in the nucleus.

- Transfer RNA (tRNA) carries amino acids to ribosomes for translation of the genetic code.

COMPARISON OF DNA AND RNA

	DNA	RNA
Type of sugar	deoxyribose	ribose
Bases	adenine, cytosine, guanine and **thymine**	adenine, cytosine, guanine and **uracil**
Number of strands in molecule	two	one
Location	only in nucleus	moves from nucleus to cytoplasm

LET'S THINK ABOUT THIS

You should be able to calculate how many molecules of an individual base are on a DNA strand when given the number of amino acids in the polypeptide and the percentage composition of any other base within the strand. Have a look at the example below.

How many guanine bases are on a DNA strand that codes for a polypeptide chain 300 amino acids long, if 20% of the DNA strand is adenine?

300 amino acids are coded for by 900 bases. If 20% of the strand is adenine, then 20% must be thymine, as they form complementary base pairs. The remaining 60% are cytosine–guanine pairs. Therefore, 30% of 900 bases are guanine = 270 bases.

PROTEIN SYNTHESIS

TRANSCRIPTION

The first stage of protein synthesis takes place in the nucleus and is called **transcription**. An mRNA molecule is produced that carries the genetic code from the DNA in the nucleus to a ribosome in the cytoplasm. Production of mRNA is essential, as DNA is too large to pass through the nuclear membrane.

The process is as follows:

1. The section of a DNA molecule carrying the code for the protein to be transcribed unwinds and 'unzips', exposing the bases.

2. mRNA nucleotides move in and form complementary base pairs with one of the DNA strands (the coding strand). Weak hydrogen bonds form. Cytosine always pairs with guanine; adenine on DNA pairs with uracil on mRNA, and thymine on DNA pairs with adenine on mRNA.

3. Strong chemical bonds form between the phosphate of one nucleotide and the ribose of the next nucleotide.

4. The weak hydrogen bonds that were holding the DNA and mRNA strands together break, allowing the mRNA to leave the nucleus and enter the cytoplasm.

5. Hydrogen bonds reform between the two DNA strands, and the DNA molecule rewinds to form a double helix.

DON'T FORGET

A triplet of bases on mRNA is called a **codon**, and codes for one amino acid.

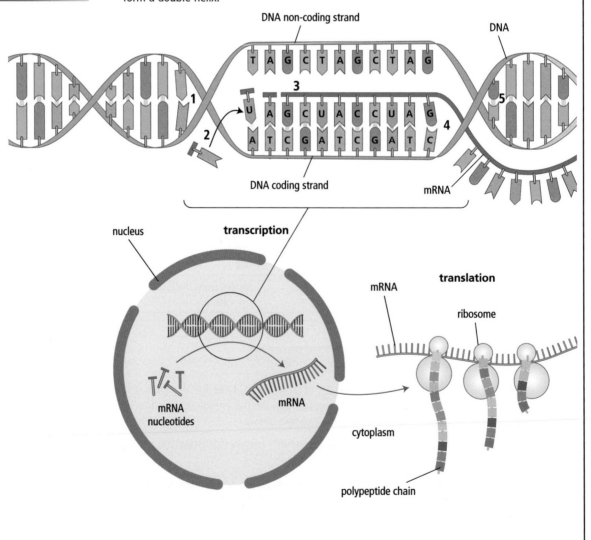

TRANSLATION

The mRNA molecule formed during transcription becomes attached to a ribosome, either lying freely in the cytoplasm or attached to the rough endoplasmic reticulum (RER). This is where **translation** occurs, in which the order of bases on the mRNA determines the order of amino acids in a specific protein. The process is as follows:

1. **Transfer RNA (tRNA)** molecules become attached to amino acid molecules in the cytoplasm. Each type of amino acid attaches to a specific tRNA molecule. As there are approximately 20 different types of amino acid, there are also approximately 20 different types of tRNA. Each type of tRNA molecule has an exposed triplet of bases called an **anticodon**.

2. Transfer RNA (tRNA) molecules in the cytoplasm transport amino acids to the ribosome.

3. The first tRNA molecule moves in to allow base pairing between the triplet of bases (anticodon) on the tRNA molecule and a triplet of bases (codon) on the mRNA strand. The codon and anticodon must be complementary to each other.

4. Another tRNA molecule carries an amino acid to the ribosome. Complementary pairing between codon and anticodon brings the amino acids in line beside each other. A peptide bond forms between the amino acids.

5. Once peptide bonds have formed between the amino acids, their position is fixed. The first tRNA molecule detaches from the mRNA and is free to collect another amino acid from the cytoplasm.

6. As translation progresses, the ribosome moves along the mRNA molecule (like a zipper moves along a zip) exposing the third codon, allowing a third tRNA molecule to bring a third amino acid into position.

7. This process is repeated until the end of the mRNA strand, when a polypeptide has been formed.

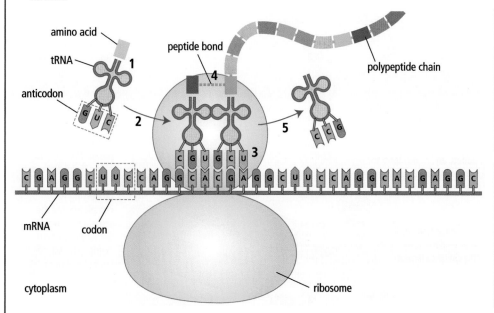

DON'T FORGET

tRNA molecules act as little buses, picking up amino acid passengers and transporting them to their destination at the ribosome.

DON'T FORGET

Translation of the genetic code is just the same as translation of foreign languages – the language of bases is converted into a language of amino acids.

Look up http://www.biotopics.co.uk/genes/trans.html
Look up http://www.artesanto.co.uk

LET'S THINK ABOUT THIS

Use the following website to practise transcription and translation of mRNA. You must know the complementary base pairs of DNA with mRNA, and mRNA with tRNA.
http://learn.genetics.utah.edu/content/begin/dna/transcribe

PROTEIN SYNTHESIS – PROCESSING AND PACKAGING

RER AND GOLGI APPARATUS

Two cell organelles are of particular importance in protein synthesis:

- The **rough endoplasmic reticulum (RER)** consists of a series of flattened membrane sacs, continuous with the nuclear membrane and with ribosomes attached to the outer membrane surface. The RER transports newly-synthesised proteins to the Golgi apparatus.

- The **Golgi apparatus** is composed of a stack of between three and seven flattened membrane sacs. Each region of the Golgi apparatus contains enzymes that alter newly-synthesised proteins by either adding bits on or chopping bits off; and vesicles move between the sacs to transfer the new proteins.

PROCESSING AND PACKAGING

Once amino acids have been assembled into a polypeptide chain at a ribosome, the polypeptide may pass into the RER and then subsequently to the Golgi apparatus for further processing and packaging. The diagram shows the sequence of events as a protein is prepared for secretion out of a cell.

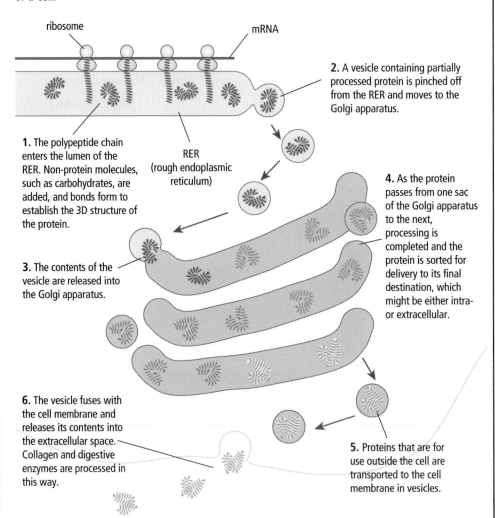

ribosome

mRNA

2. A vesicle containing partially processed protein is pinched off from the RER and moves to the Golgi apparatus.

1. The polypeptide chain enters the lumen of the RER. Non-protein molecules, such as carbohydrates, are added, and bonds form to establish the 3D structure of the protein.

RER (rough endoplasmic reticulum)

4. As the protein passes from one sac of the Golgi apparatus to the next, processing is completed and the protein is sorted for delivery to its final destination, which might be either intra- or extracellular.

3. The contents of the vesicle are released into the Golgi apparatus.

6. The vesicle fuses with the cell membrane and releases its contents into the extracellular space. Collagen and digestive enzymes are processed in this way.

5. Proteins that are for use outside the cell are transported to the cell membrane in vesicles.

 Look up http://www.johnkyrk.com/golgiAlone.html

 **DON'T FORGET**

The Golgi apparatus works just like a sorting office – packaging, labelling and sorting the new proteins before sending them on to the place where they are needed.

SUMMARY OF PROTEIN SYNTHESIS

mRNA nucleotides are assembled using a DNA template. The language of the code as a sequence of bases on DNA is retained as a sequence of bases on mRNA. It is simply transcribed (just like copying this paragraph onto another sheet of paper).

Proteins are passed through RER and on to the Golgi apparatus, where they are packaged before use and may have non-protein molecules such as carbohydrates added to them. Vesicles transport the protein between RER and Golgi apparatus.

processing and packaging

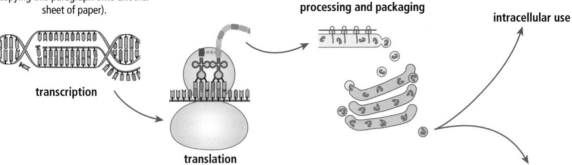

intracellular use

transcription

translation

mRNA moves to a ribosome in the cytoplasm. tRNA molecules pick up amino acids and transport them to the ribosome. Each tRNA anticodon binds temporarily with the corresponding codon on the mRNA. This allows the amino acids to be assembled in the correct sequence for a particular protein. Once peptide bonds have formed between the amino acids, each tRNA leaves the ribosome and goes back to pick up another amino acid. The language of the code has been changed from a sequence of bases into a sequence of amino acids. It has been translated!

extracellular use

Secretory vesicles leave the Golgi apparatus and move to the cell surface. The vesicle membrane fuses with the cell membrane to release the contents into the extracellular space. Hormones and digestive enzymes are processed in this way.

LET'S THINK ABOUT THIS

When given appropriate information, you should be able to work out corresponding sequences of DNA, RNA and peptide chains. The table below shows the mRNA codons for some amino acids.

First position	Second position				Third position
	A	**U**	**G**	**C**	
A	lycine asparagine	methionine/start isoleucine	arginine serine	threonine threonine	G U
C	glutamine histidine	leucine leucine	arginine arginine	proline proline	G U

(a) A section of DNA has the following bases:

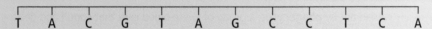

T   A   C   G   T   A   G   C   C   T   C   A

Identify the codons of the mRNA and the anticodons of the tRNA that would be used in protein synthesis from this section of DNA. Use the table of mRNA codons to identify the amino acid sequence that would be produced.

(b) From the table, what are the mRNA codons for (i) lycine and (ii) isoleucine?

(c) What is the tRNA anticodon for the amino acid arginine?

LET'S THINK ABOUT THIS

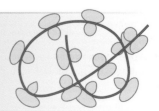

Ribosomes that assemble proteins for use outside the cell are always attached to the RER. However, ribosomes can also be found free within the cytoplasm. Groups of ribosomes become attached to one mRNA strand, all translating the code at the same time, very quickly producing many copies of the same polypeptide for use inside the cell.

VIRUSES AND DEFENCE MECHANISMS IN ANIMALS

VIRUSES

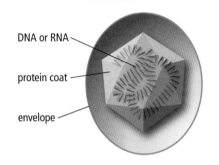

DNA or RNA

protein coat

envelope

Viruses are very small, ranging in size from 10–30 nm. They contain **nucleic acid** (either **DNA or RNA**) surrounded by a **protein coat**. Viruses only replicate by invading another cell, altering the host cell's metabolism to allow replication of viral DNA and RNA, so that new viruses can be produced. The choice of host cell is often very specific – the polio virus attacks neurones, and the hepatitis virus attacks liver cells. Viruses can remain dormant within the host cell for considerable periods of time before becoming active.

Viral replication

Viral nucleic acid is replicated using the host cell's nucleotides and ATP for energy.		Each copy of viral DNA becomes enclosed in a protein coat.

Virus attaches to the host cell.			

attachment ➝ penetration ➝ DNA replication ➝ protein synthesis ➝ assembly ➝ release

The virus releases its nucleic acid into the host cell. The virus alters the metabolism of the cell so it stops performing its normal function.

Using the new copies of the viral nucleic acid, viral mRNA is transcribed, allowing protein coats to be made from the host cell's amino acids and ATP.

The host cell bursts, releasing large numbers of viruses.

Look up www.whfreeman.com/kuby/content/anm/kb03an01.htm

PHAGOCYTOSIS

Phagocytosis is carried out by **phagocytes**, a group of white blood cells that engulf and destroy foreign bodies such as bacteria.

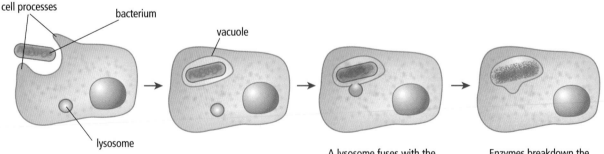

cell processes

bacterium

vacuole

lysosome

Cell processes surround the bacterium, enclosing it in a vacuole.

The vacuole moves into the cell.

A lysosome fuses with the vacuole, releasing digestive enzymes.

Enzymes breakdown the bacterium. Products may be reused by the cell.

ANTIBODY PRODUCTION

When a virus invades the body, molecules on its surface act as **antigens** that are recognised as foreign by white blood cells called **lymphocytes**. In response, the lymphocytes multiply so that they can produce a large quantity of **antibodies**. Antibodies are Y-shaped molecules with two receptor sites, one on each arm. The shape of this binding site is **specific** to allow attachment only to the antigen that had initially been recognised. Once the antibody locks on to the antigen, both are removed by phagocytes.

DON'T FORGET

Antibodies are specific – they act on only one antigen.

Look up
www.geocities.com/victor
.castilla/humoral.html

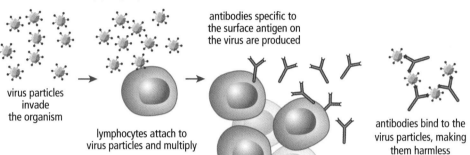

virus particles invade the organism

lymphocytes attach to virus particles and multiply

antibodies specific to the surface antigen on the virus are produced

antibodies bind to the virus particles, making them harmless

TISSUE REJECTION

The cells of a transplanted organ are regarded as **foreign** by the recipient's immune system, because the organ's cell surfaces carry non-self antigens. This causes lymphocytes to attack the organ as though it is a disease-causing pathogen, and organ failure can result. To try to prevent rejection, the organ is **tissue typed** to match the donor and recipient as closely as possible, and the recipient receives **immunosuppressant drugs**.

PRIMARY AND SECONDARY RESPONSE

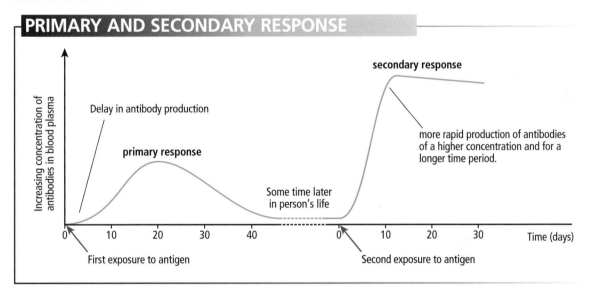

LET'S THINK ABOUT THIS

The ways by which an animal can gain antibody protection are shown below.

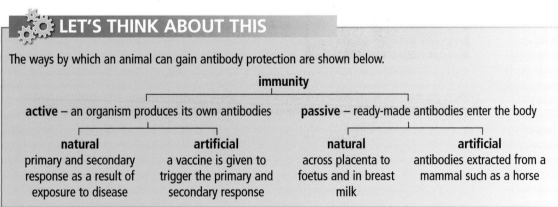

DEFENCE MECHANISMS IN PLANTS

Plants protect themselves in one of two ways:

- by producing toxic compounds
- by establishing a physical barrier that isolates the damaged tissue and the attacking organism.

PRODUCTION OF TOXIC COMPOUNDS

Chemical produced	How it protects the plant
tannins	Tannins act as enzyme inhibitors that interfere with cell metabolism. In some fruits and in the leaves of plants such as tea and eucalyptus, tannins make the tissue less palatable to deter grazing animals.
cyanide	Leaf cells produce hydrogen cyanide which acts by blocking the cytochrome system of a grazing animal's cells, preventing ATP from being produced in aerobic respiration. This deters the animal from grazing on the plant. Production of hydrogen cyanide (cyanogenesis) only begins after damage to the plant has occurred. Plants that produce hydrogen cyanide are said to be **cyanogenic**.
nicotine	Nicotine is a strong toxin that is made in the roots of plants and accumulates in the leaves. It is found in leaves of the tobacco family. Nicotine discourages insect attack.

ISOLATION OF THE INJURED AREA

Barrier	How this prevents further attack
resin 	Resin is a sticky substance that is produced by cells at a wound surface. The resin acts by isolating invading pathogens, preventing them from spreading throughout the plant.

Look up http://www.biology-online.org/1/12_cell_defense_3.htm

LET'S THINK ABOUT THIS

You should be prepared to write an essay on cellular defence mechanisms. A sample answer is shown below.

Give an account of cellular defence mechanisms in animals and plants.

In animals, two types of white blood cell are important in defence against invading microbes: phagocytes and lymphocytes.

Phagocytes send out cell processes that extend around the invading microbe, enclosing it in a vacuole. Lysosomes containing digestive enzymes then fuse with the vacuole. The digestive enzymes break down and destroy the microbe. Lymphocytes detect antigens on the surface of the microbe. They produce antibodies. These are Y-shaped protein molecules that can bind to specific antigens. Once an antibody has attached to the microbe, the microbe is rendered harmless.

Plants have two different groups of defence mechanisms: production of toxic chemicals, and isolation of damaged tissues. Toxic chemicals include tannins, cyanide and nicotine. Tannins are common in some fruits and in leaves of plants such as tea and eucalyptus. They act as enzyme inhibitors that interfere with cell metabolism. In leaves, they provide defence by making the leaves less palatable to grazing animals. Some plants are cyanogenic and produce hydrogen cyanide. This toxin blocks the cytochrome system of a grazing animal's cells. Cyanogenic plants only start to produce hydrogen cyanide after a plant has been grazed on. As a result, older plants are less likely to be grazed on than young plants. Nicotine is found in plants such as tobacco. It is a strong toxin that acts to discourage attack by insects.

Plants also defend against attack by producing resin. This sticky substance is produced around the site of damage. It isolates invading pathogens in the damaged tissue, acting as a physical barrier to prevent them from spreading to other areas of the plant.

VARIATION

MEIOSIS

Meiosis is the type of cell division that results in the production of sex cells (**gametes**) and takes place in the testes and the ovaries. The events of meiosis are important for two reasons:

1. In sexual reproduction, two **gametes** fuse to produce a **zygote**. To maintain the correct number of chromosomes in the new individual, the gametes must contain half the number of chromosomes found in all other cells of the organism. The gametes are said to be **haploid** (contain one set of chromosomes); all other cells are **diploid** (contain two sets of chromosomes).

2. Meiosis allows genetic **variation** to be introduced. For any species, variation is the key to survival, as the differences it produces may allow some individuals to survive environmental change. Over a period of time, variation can lead to the **evolution** of a new species.

Meiosis consists of two divisions:

- The **first meiotic division (meiosis I)**, which reduces the number of chromosomes.

- The **second meiotic division (meiosis II)**, which reduces the quantity of DNA on each chromosome.

DON'T FORGET

Variation is introduced during crossing over and independent assortment.

Meiosis I

During **interphase**, chromosomes of the **gamete mother cell** are long, thin and invisible in the cell. Towards the end of this stage, the DNA replicates.

In **prophase I**, the chromosomes become shorter and fatter (**condensed**) to form the characteristic X-shape. Each chromosome is composed of two **chromatids** held together by a **centromere**.

Homologous chromosomes line up together on a gene-for-gene basis. Adjacent chromatids join at several points along their length, called **chiasmata** (singular = **chiasma**), and exchange segments in a process known as **crossing over**. Once crossing over is complete, the chromatids separate.

At the end of prophase I, the nuclear membrane breaks down.

The spindle invades the central region of the cell in **metaphase I**, and homologous chromosomes line up together on either side of the equator. The orientation of each chromosome pair is random to all other pairs (**independent assortment**).

During **anaphase I**, homologous pairs are pulled apart to opposite poles of the cell. Because the centromere has not been copied, each chromosome is still double-stranded.

In **telophase I**, nuclear membranes form and the cytoplasm divides to produce two daughter cells, each with half the number of chromosomes of the gamete mother cell. But a second meiotic division is now necessary to reduce the number of chromatids on each chromosome.

contd

MEIOSIS contd

Meiosis II

In **prophase II**, a new spindle forms in the cytoplasm before the nuclear membrane breaks down.

Chromosomes line up on the equator during **metaphase II**. At the end of this stage, the centromere replicates to allow chromatids to separate.

During **anaphase II**, chromatids are pulled to opposite poles of the cell, forming single-stranded chromosomes.

New nuclear membranes form during **telophase II**, and the cytoplasm divides by **cytokinesis**. At the end of telophase II, each gamete mother cell has produced four haploid daughter cells, each containing half the number of single-stranded chromosomes.

DON'T FORGET

Meiosis II involves the same stages as mitosis.

Look up
www.cellsalive.com/meiosis.htm

CROSSING OVER

Crossing over occurs during prophase I. It is important in introducing variation, as different combinations of alleles are produced in the gametes.

homologous chromosomes pair up

chiasmata form allowing crossing over

recombination of alleles occurs

INDEPENDENT ASSORTMENT

During metaphase I, homologous chromosomes line up on either side of the equator. The side of the equator that each chromosome ends up on is entirely random and independent of all the other chromosome pairs. By increasing the possible combinations of chromosomes in any gamete, variation is again increased.

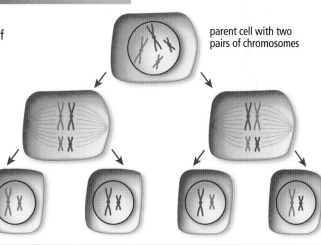

parent cell with two pairs of chromosomes

possible alignments at metaphase I

four possible combinations of chromosomes at telophase I

LET'S THINK ABOUT THIS

Try drawing out all the possible gametes that could result from independent assortment when the gamete mother cell contains a diploid number of six chromosomes. (Hint: there are eight possible combinations of chromosomes.)

gamete mother cell

THE DIHYBRID CROSS AND LINKED GENES

THE DIHYBRID CROSS

In a dihybrid cross, the inheritance of two genes for different characteristics is considered. Let's look at the inheritance of plant height (gene A) and flower colour (gene B) in pea plants.

parents (P)	phenotype	true breeding for tall, pink flower	X	true breeding for dwarf, white flower
genotype		AABB		aabb
gametes		all AB		all ab
F_1	genotype		all AaBb	
	phenotype		all tall, pink flower	

F_1 cross   AaBb   X   AaBb

F_1 gametes   AB or Ab or aB or ab   AB or Ab or aB or ab

F_2 punnet square

	AB	Ab	aB	ab
AB	AABB	AABb	AaBB	AaBb
Ab	AABb	AAbb	AaBb	Aabb
aB	AaBB	AaBb	aaBB	aaBb
ab	AaBb	Aabb	aaBb	aabb

Possible F_2 genotype	Expected number	Possible F_2 phenotype	
AABB	1	tall, pink flower	parental
AABb	2	tall, pink flower	parental
AAbb	1	tall, white flower	recombinant
AaBB	2	tall, pink flower	parental
AaBb	4	tall, pink flower	parental
Aabb	2	tall, white flower	recombinant
aaBB	1	dwarf, pink flower	recombinant
aaBb	2	dwarf, pink flower	recombinant
aabb	1	dwarf, white flower	parental

Add up the expected number of F_2 offspring with each phenotype to get the ratio:
9 tall, pink: **3** tall, white: **3** dwarf, pink: **1** dwarf, white

The expected ratio of 9:3:3:1 may not occur in the actual population for three reasons:

1. The results were not reliable due to a small population size.

2. The genes under examination are carried on the same chromosome (**linked genes**).

3. Some gametes were not viable.

LINKED GENES

Genes that are located on the same chromosome form **linkage groups**. You would expect the alleles of linked genes to stay together throughout meiosis. For example, in *Drosophila*, if a grey body allele is on the same chromosome as a normal wing allele, the gamete inherits both the grey body and normal wing alleles. However, crossing over during meiosis I can separate the genes, forming new combinations of alleles in the gametes (**recombinant gametes**).

contd

LINKED GENES contd

Example of a linked gene cross

Consider a dihybrid cross in sweet pea plants, where the genes are located on the same chromosome (linked genes). The allele for purple flowers (P) is dominant to red flowers (p); long pollen grain (L) is dominant to round pollen grain (l). With no crossing over, a cross between a true-breeding purple flower, long pollen grain plant and a true-breeding red flower, round pollen grain plant would be as in this diagram:

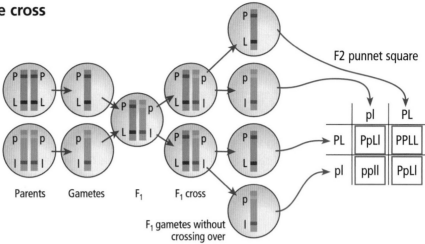

The F$_2$ offspring would have the expected genotypes and phenotypes shown in the table, giving an expected phenotype ratio of 3:1.

Genotype	Number	Phenotype
PPLL	1	purple flower, long pollen grain
PpLl	2	
ppll	1	red flower, round pollen grain

Now consider the same cross but this time **with crossing over**, where linkage groups are broken.

Crossing over has produced F$_2$ offspring in which the combination of phenotypes has been altered, permitting purple flower and round pollen, and red flower and long pollen. These are **recombinant offspring**. Crossing over of linked genes results in a phenotype ratio which is different to the expected 3:1 ratio of linked genes.

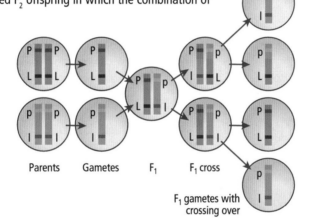

F2 punnet square

	Pl	pL	PL	pl
Pl	PPll	PpLl	PPLl	Ppll
pL	PpLl	ppLL	PpLL	ppLl
PL	PPLl	PpLL	PPLL	PpLl
pl	Ppll	ppLl	PpLl	ppll

Look up www.biostudio.com/d_%20Meiotic%20Recombination%20Between %20Linked%20Genes.htm

LET'S THINK ABOUT THIS

You should be able to determine the order of four genes on a chromosome, given the percentage recombination frequency of the genes. To do this, remember that the larger the percentage recombination frequency, the further apart the genes.

Gene pair	A and B	C and B	C and D	A and C	B and D
% recombination frequency	33	37	31	4	6

C	A	D	B

C and B have the greatest percentage recombination frequency, so must be furthest apart. A and B have a larger recombination frequency than B and D. Therefore, D must be next to B, with A between C and D.

SEX DETERMINATION AND SEX-LINKED GENES

SEX DETERMINATION

In humans, there are 23 pairs of chromosomes. Only one of these pairs (the **sex chromosomes**, **X** and **Y**) is involved with sex determination. Females have two X chromosomes and males have an X chromosome and a Y chromosome. The diagram below shows how sex is determined at fertilisation.

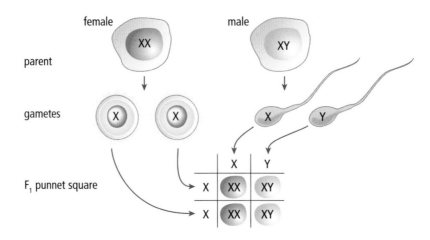

DON'T FORGET

The expected male:female ratio is 1:1.

The female can only produce gametes containing an X chromosome, but the male produces sperm that carry either an X chromosome or a Y chromosome. The male therefore determines the sex of the offspring.

SEX LINKAGE

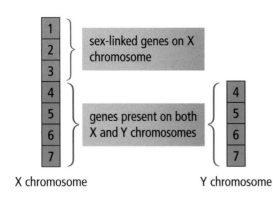

X chromosome   Y chromosome

Although the X and Y chromosomes are homologous chromosomes, the difference in their size means that some genes are found on both chromosomes and others are found only on the larger X chromosome.

In the diagram, genes 1–3 are found only on the X chromosome and are called **sex-linked genes**. A male inherits one X and one Y chromosome – and so, for the sex-linked genes, he will only have one allele (on the X chromosome). This means that the characteristic of the gene on the X chromosome will be expressed, no matter if a dominant or recessive allele has been inherited (remember that, in non-sex-linked genes, two recessive alleles must be present to display the recessive phenotype).

EXAMPLES OF SEX LINKAGE

Red–green colour blindness in humans

Normal vision (N) is dominant and red–green colour blindness (n) is recessive. This gene is a sex-linked gene, which means it is located on the X chromosome.

Consider a cross between a female homozygous for normal vision and a colour-blind male.

contd

EXAMPLES OF SEX LINKAGE contd

parents	$X^N X^N$	×	$X^n Y$
gametes	all X^N		X^n or Y
F_1	$X^N X^n$	or	$X^N Y$
	carrier female		normal male

The F_1 female has inherited the colour-blindness gene from her father but, as she has also inherited a dominant gene for normal sight from her mother, she has normal vision. She is, however, a **carrier** and could pass on the colour-blindness gene to her offspring.

Red–green colour blindness is quite rare in females because they have two X chromosomes and so would have to inherit two recessive alleles. In males, inheriting an X chromosome with a recessive gene will give rise to colour blindness.

The inheritance of colour blindness can be traced through generations using a family tree.

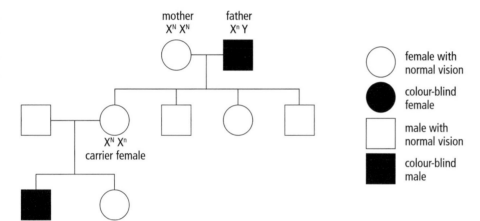

mother
$X^N X^N$

father
$X^n Y$

$X^N X^n$
carrier female

female with normal vision

colour-blind female

male with normal vision

colour-blind male

This family tree shows that sex-linked characteristics can skip generations. The male parent is affected but none of his children is colour blind. One of his daughters is a carrier, and she has passed on the colour-blindness gene to her son. More males than females are affected.

DON'T FORGET

Sex-linked genes are always indicated using the letters X and Y in the genotype.

DON'T FORGET

If representing genes with letters that are very similar in upper and lower case, for example C and c, make sure they are clearly different sizes so that examiners can see which allele you are referring to.

LET'S THINK ABOUT THIS

You should be able to work out the theoretical outcomes of crosses involving sex-linked genes.

1. What are the possible F_1 genotypes and phenotypes from the following crosses?

	Female		Male	
	genotype	phenotype	genotype	phenotype
Cross 1	$X^N X^n$	carrier	$X^n Y$	colour blind
Cross 2	$X^N X^n$	carrier	$X^N Y$	normal vision

2. Haemophilia is another sex-linked condition. Write down all the possible genotypes and phenotypes associated with this gene.

LET'S THINK ABOUT THIS

Haemophilia first appeared in the British Royal Family with Queen Victoria's children, although there was no history of the condition in the families of either Queen Victoria or Prince Albert. This is explained by the presence of a mutant recessive gene on one of Queen Victoria's X chromosomes.

MUTATIONS – CHANGES TO GENE STRUCTURE AND CHROMOSOME STRUCTURE

MUTAGENIC AGENTS

A **mutation** is an alteration to the genetic information in a cell. **Random** mutations occur naturally within a population, usually at a very **low frequency**, and introduce variation to the species. However, some agents cause the rate of mutation to increase. Examples of **mutagenic agents** include irradiation by ultra-violet light, gamma rays, or X-rays; and a variety of chemicals such as those in mustard gas and cigarette smoke.

GENE MUTATIONS

A gene is a segment of a DNA molecule that carries the code for one protein. The code takes the form of a sequence of bases, with a triplet of bases coding for one amino acid (see page 26). If the sequence of bases is altered, the corresponding sequence of amino acids may change, possibly altering the protein produced. There are four types of gene mutation:

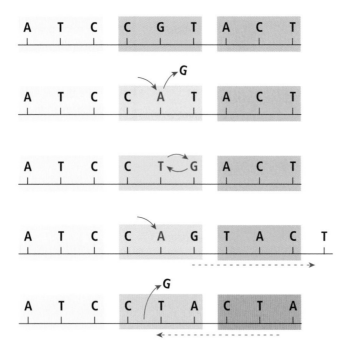

normal strand of DNA

Substitution – one of the bases is replaced by a different base (here A replaces G). This is a **point mutation**, as it only changes the amino acid coded for by the affected triplet.

Inversion – some of the bases swap position (here G, T has been inverted to give T, G). This is another **point mutation**, as only the amino acid coded for by the affected triplet is altered.

Insertion – an extra base (A) has been inserted into the sequence, moving the bases after the insertion one place to the right. This is a **frame-shift mutation**, as every amino acid after the point of insertion is altered.

Deletion – a base has been removed (G) from the sequence, shifting the bases after the deletion to the left. This is another **frame-shift mutation**, as every amino acid after the point of deletion is altered.

 Look up www.bbc.co.uk/Scotland/learning/bitesize/higher/biology/genetics _adaptation/mutations1_rev1.shtml

As point mutations usually alter just one amino acid, the effect on any protein produced is generally minor. However, sometimes the amino acid that is altered is in a key position for protein function. An example of this is sickle-cell anaemia, where normal haemoglobin is replaced by haemoglobin S.

Frame-shift mutations always cause major changes to the amino acid sequence and therefore greatly alter the protein that is synthesised. Frame-shift mutations frequently result in non-viable gametes.

CHANGES TO CHROMOSOME STRUCTURE

Sometimes, crossing over can result in mutations. When segments of homologous DNA break and move between chromosomes, the rearranged structure can end up with the wrong number of genes. These changes can be divided into four types of mutation:

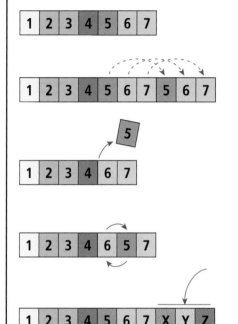

Normal gene sequence

Duplication – the chromosome has gained several genes (5, 6, 7) from its homologous partner. This may be beneficial to the evolution of the species if extra gene copies give the organism an advantage.

Deletion – one or more genes (in this case gene 5) have been lost from the chromosome. Loss of a gene can be serious, as important proteins for normal metabolism may not be produced.

Inversion – a segment of the chromosome (5, 6) has been turned around. All genes are still present, but in the wrong order. Non-viable gametes are often produced.

Translocation – part of a non-homologous chromosome (X, Y, Z) has become attached. Non-viable gametes are usually produced.

 Look up www.bbc.co.uk/Scotland/learning/bitesize/higher/biology/genetics _adaptation/mutations1_rev1.shtml

LET'S THINK ABOUT THIS

It is very easy to get muddled up between the different types of mutation, particularly as deletion and inversion are types of both gene and chromosome mutations. You must pay particular attention to this when answering exam questions on this topic. It helps in essays to use letters to represent base sequences in gene mutations, and numbers to represent whole genes in structural changes to chromosomes.

1. The correct base sequence on a DNA strand is shown below:

| T | G | A | A | C | G | A | C | T |

Using the information above, identify the following gene mutations.

(a) | T | G | A | A | C | G | C | C | T |

(b) | T | G | A | C | G | A | C | T |

(c) | A | G | T | A | C | G | A | C | T |

2. A chromosome has undergone a deletion mutation. Describe the events that may have produced this mutation.

MUTATIONS – CHANGES TO CHROMOSOME NUMBER

During cell division, the spindle moves chromosomes from the equator of the cell to the poles. Sometimes, the spindle doesn't function properly and chromosomes are not pulled apart, ending up in the wrong cell. This is called **non-disjunction** and results in the production of gametes that have either extra or missing chromosomes. The diagram below shows non-disjunction in meiosis I.

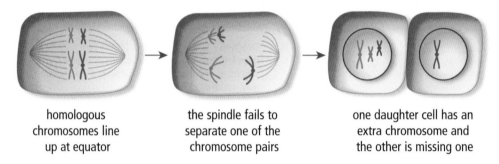

| homologous chromosomes line up at equator | the spindle fails to separate one of the chromosome pairs | one daughter cell has an extra chromosome and the other is missing one |

Sometimes, complete sets of chromosomes fail to separate and one daughter cell gains an extra set of chromosomes. This condition, called **polyploidy**, happens when there is complete breakdown of the cell spindle. Polyploidy is extremely rare in animals but is more frequent in plants.

NON-DISJUNCTION IN ANIMALS

Non-disjunction can affect non-sex chromosomes (**autosomes**) and the sex chromosomes. Non-disjunction of chromosome number 21 causes **Down's Syndrome**. In Down's Syndrome, the individual has an extra chromosome number 21 and a characteristic appearance, including prominent, slanted eyelids and a short nose. Mental disability also results.

When sex chromosomes are absent from one gamete and an X chromosome is present in the other gamete, a condition known as Turner's Syndrome results. Individuals are infertile females.

Kleinfelter's Syndrome is also caused by non-disjunction of sex chromosomes. This time, an egg containing two X chromosomes is fertilised by a sperm containing a Y chromosome to produce an infertile male.

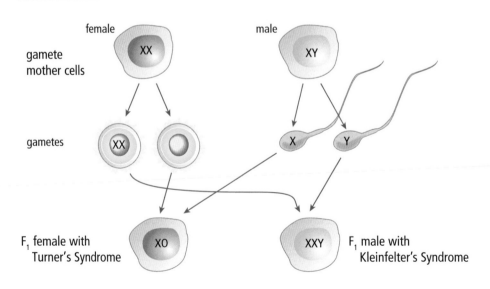

NON-DISJUNCTION IN PLANTS

Polyploidy in plants can be extremely useful both to the plant and to humans, and can be induced experimentally by heat or cold shock, or by exposure to some chemicals, such as **cholchicine**. Plants with cells containing three or more complete sets of chromosomes often grow larger than those plants with the normal diploid complement of chromosomes. This is of economic importance in cultivated plants, such as wheat and coffee, whose polyploid varieties produce a much higher yield than the diploid variety.

When polyploidy is produced by crossing two different varieties of plant, the resulting hybrid often shows improved fertility, growth or resistance to disease. These improvements in hybrids are collectively called **hybrid vigour**.

The diagram below shows how polyploidy can occur in plants.

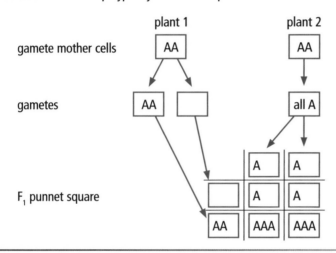

LET'S THINK ABOUT THIS

1. An individual has body cells containing only 21 autosomes and two sex chromosomes. Describe how this mutation could have arisen.

2. Explain why polyploidy in plants may be of use to humans.

LET'S THINK ABOUT THIS

The incidence of Down's Syndrome increases with maternal age. This increase in frequency is related to the length of time that the spindle is in position, because the gamete mother cells in the ovary enter metaphase I when the woman is still a foetus herself. Meiosis I does not continue until some point after puberty, when that particular gamete mother cell is selected for further development in the ovarian cycle. It may be several decades after forming that the spindle pulls on the chromosomes to separate them.

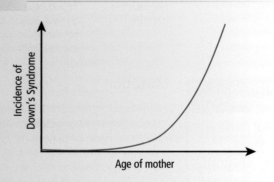

NATURAL SELECTION – SPECIATION AND ADAPTIVE RADIATION

SPECIES

A **species** is a group of organisms that can interbreed to produce fertile young. Interbreeding sometimes occurs between species. However, the offspring are infertile.

SPECIATION

Speciation is the process that results in the formation of a **new species**. It depends on three factors:

1. isolation　　**2.** mutation　　**3.** natural selection.

Isolation

Different groups within a species (**subpopulations**) can become isolated from each other so that interbreeding and, therefore, mixing of genes between subpopulations does not take place. There are several ways that subpopulations can become isolated.

geographical isolation	When new mountain ranges, deserts, rivers or seas form, subpopulations can become separated, being unable to cross the physical barrier to interbreed.

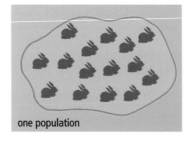

one population

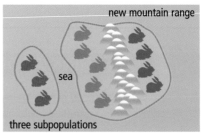

new mountain range

sea

three subpopulations

ecological isolation	If habitats start to change, the preferred habitat of one group can become hostile to other subpopulations. Changes can include differences in moisture level, temperature and pH.
reproductive isolation	Reproductive differences between the subpopulations can prevent interbreeding. For example, subpopulations flower at different times; there may be a lack of attraction between males and females of different subgroups; or there may be physical incompatibility of genital organs between subgroups.

Mutation

Random mutation occurs in all populations and is the means of introducing **variation** into a species. When subpopulations cannot interbreed – because an isolation barrier exists – new genes do not spread between groups. Sometimes, a mutation occurs that benefits an individual, making it more likely that the individual survives, breeds successfully and produces more offspring. The mutation gives the individual a **reproductive advantage**.

Natural selection

Individuals with characteristics that make them better suited to their environment are more likely to breed. This is **survival of the fittest** and is the basis for **natural selection**. The term 'fittest' might refer to the fastest prey organism that can outrun a predator, or maybe a plant that has become drought-resistant in a dry habitat. Each of these new characteristics is coded for by a mutant gene or by a new combination of genes. Over a long period of time, the number of mutations that occurs in a subpopulation continues to increase. Eventually, the genetic make-up (**gene pool**) of the subpopulations diverges so far that individuals become genetically too different to interbreed.

DON'T FORGET

Remember that members of the same species become isolated BEFORE new species form.

Look up
www.pbs.org/wgbh/
evolution/sex/guppy/
index.html

ADAPTIVE RADIATION

Adaptive radiation is the evolution of several new species from a relatively unspecialised **common ancestor**. The new species show phenotypes that are specialised to allow organisms to exploit different habitats or food sources (**niches**). Opportunities for adaptive radiation can occur when either:

- a radical change occurs within a colonised environment, or
- isolated ecosystems are colonised by new species.

Marsupials

Marsupials are a group of mammals that are found only in Australia, which was – at one time – attached by land bridges to other continents. When the continents fragmented and the land bridges disappeared, a geological barrier (the ocean) formed, isolating Australia's primitive mammals from those on other land masses. A common mammalian ancestor is thought to have given rise to all marsupials – and, as subpopulations spread out on the continent, natural selection allowed different species to evolve to occupy different niches.

Darwin's Finches

A small flock of finches migrated 600 miles from the South American mainland to the isolated Galapagos Islands. Once there, the birds scattered across the different volcanic islands and, in the absence of competition, increased in number. Subpopulations became isolated from other groups and adapted to take advantage of different types of food. From these subpopulations, 14 species of finch have evolved, distinguished by the shape and size of their beak, and by behaviour.

eats insects in trees – probing beak

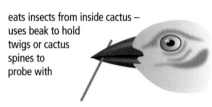

eats insects from inside cactus – uses beak to hold twigs or cactus spines to probe with

eats insects in trees – grasping beak

eats cactus seeds – crushing beak

 Look up www.pbs.org/wgbh/evolution/darwin/origin/index.html

Industrial melanism in peppered moths

Although evolution usually takes place over a long period of time, sometimes a radical environmental change can drive evolution at a much faster rate. The Industrial Revolution in Britain caused a rapid increase in air pollution, which resulted in the death of tree lichens and blackening of tree bark by soot. These factors had an effect on the distribution of peppered moths. In industrial areas, the darker melanic variety was less visible against the dark tree bark and was less likely to be eaten by predators, giving this form a reproductive advantage. However in non-industrial areas, the paler tree bark provides camouflage for the lighter speckled variety.

 DON'T FORGET

The melanic form was the product of normal variation, not the result of pollution.

 Look up www.echalk.co.uk/Science/tasters/biology/taster.htm

LET'S THINK ABOUT THIS

Antibiotic-resistant strains of bacteria have evolved in the context of the widespread use of antibiotics. Mutations give resistance, and subpopulations survive treatment with antibiotics to multiply, while the normal variety is killed. MRSA and *Clostridium difficile* are drug-resistant bacteria.

Try to make a bullet-point list of the stages involved in speciation and adaptive radiation, highlighting in your answer the existence of a common ancestor and the opportunity to exploit a new niche.

ARTIFICIAL SELECTION 1

Humans use artificial selection to make improvements in the characteristics of organisms. There are several ways that this can be done:

- **selective breeding**
- **hybridisation**
- **somatic fusion**
- **genetic engineering.**

SELECTIVE BREEDING

> **DON'T FORGET**
>
> Humans can select individuals with certain characteristics for breeding, but the variation within a species arises naturally.

Parent organisms with the desired characteristics are **deliberately selected** by humans and used for breeding. In each new generation, the offspring that inherit the desired characteristic are selected and the process is repeated. The quality of the characteristic gradually improves.

Modern varieties that have been selectively bred sometimes look very different from the ancestor species. Broccoli, cabbage, kale, brussels sprouts and cauliflower were all bred from the same ancestor – wild mustard. To produce the different species, wild mustard plants with particular characteristics were chosen for breeding. For example, to arrive at the modern brussels sprout plant, ancestors with strong lateral buds would have been selected in every generation, whereas plants with a strong terminal bud would have been used to obtain the characteristics of a cabbage.

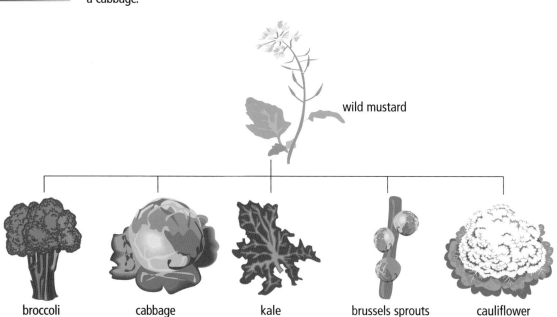

wild mustard

broccoli   cabbage   kale   brussels sprouts   cauliflower

Selective breeding has been used to produce:

- crop plants that give a **higher yield** (wheat and corn)
- cattle that produce more milk or give increased meat production
- sheep that provide higher wool yield.

Selective breeding has two main disadvantages: it takes a long time, and it sometimes leads to unwanted characteristics being passed on with or instead of the selected characteristic.

HYBRIDISATION

If two species have different characteristics that would be desirable in the same organism, they can be bred together (**hybridisation**) to produce a **hybrid** organism. The increased gene pool that is available during hybridisation often results in a hybrid that has improved growth over the original varieties (**hybrid vigour**).

SOMATIC FUSION

In the case of plant species that are sexually incompatible and cannot interbreed, **somatic fusion** is used. This process involves removing the cell wall from two cells (not gametes) to produce **protoplasts**, which are then fused together. By fusing two diploid cells, a polyploid cell (with four sets of chromosomes) is produced that can be cultured to form new hybrid plants which display the characteristics of both parents.

This technique has been used in potato plants to produce varieties that give a high yield and are resistant to diseases.

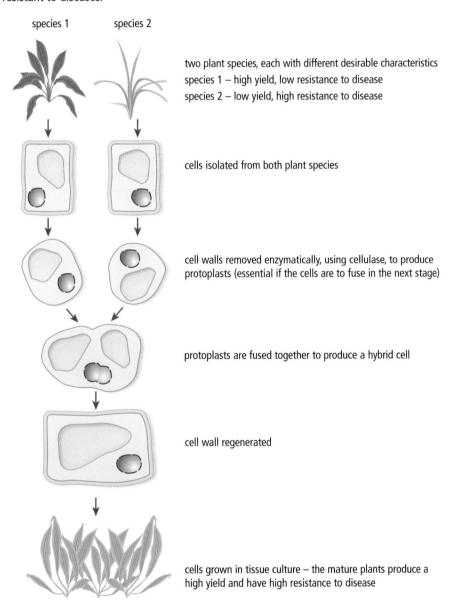

species 1   species 2

two plant species, each with different desirable characteristics
species 1 – high yield, low resistance to disease
species 2 – low yield, high resistance to disease

cells isolated from both plant species

cell walls removed enzymatically, using cellulase, to produce protoplasts (essential if the cells are to fuse in the next stage)

protoplasts are fused together to produce a hybrid cell

cell wall regenerated

cells grown in tissue culture – the mature plants produce a high yield and have high resistance to disease

LET'S THINK ABOUT THIS

Dogs have been bred selectively to produce the many different varieties that exist today. To do this, generations of animals have been inbred. Inbreeding increases the chances of recessive alleles being expressed in the phenotype (compared with hybridisation, in which recessive characteristics are seen less often). As a result, some breeds of dog are prone to particular conditions – Dalmatians are prone to deafness, and cocker spaniels are prone to kidney disease.

ARTIFICIAL SELECTION 2

GENETIC ENGINEERING

Genes from one species can be 'cut out and pasted' into the **genome** of another unrelated species using genetic engineering. This can transfer characteristics that make the genetically-altered organism more useful or more productive. Genetic engineering has the following advantages over selective breeding:

- Accuracy – the chosen characteristic is transferred, and unwanted characteristics are kept to a minimum.

- Speed – the desired characteristic can be expressed in one generation.

- Flexibility – new combinations of characteristics that would not otherwise be possible can be achieved.

Although it is possible to insert genes into many different types of cell, bacteria are frequently used in genetic engineering, as it is relatively easy to isolate and transfer bacterial DNA. For example, genetically-engineered bacteria containing human genes are used to synthesise large quantities of human proteins, such as insulin and human growth hormone.

The diagram below summarises the processes involved.

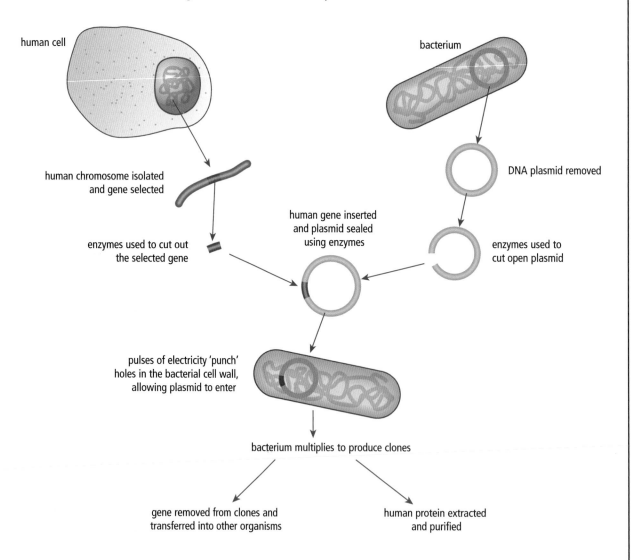

human cell

bacterium

human chromosome isolated
and gene selected

DNA plasmid removed

enzymes used to cut out
the selected gene

human gene inserted
and plasmid sealed
using enzymes

enzymes used to
cut open plasmid

pulses of electricity 'punch'
holes in the bacterial cell wall,
allowing plasmid to enter

bacterium multiplies to produce clones

gene removed from clones and
transferred into other organisms

human protein extracted
and purified

contd

GENETIC ENGINEERING contd

Two types of enzyme are important for this process:

1. **Endonucleases (restriction enzymes)** – these cut genes out of DNA molecules and can be used to open bacterial plasmids, making sure that the ends of the DNA fragments are complementary.

2. **Ligase** is used to seal the new genes into the bacterial plasmid.

DON'T FORGET

Think of endonucleases as scissors and ligase as glue.

Uses of genetic engineering

The table below shows some of the species that are being used in genetic engineering at the present time.

Species	New characteristic	Benefit to humans
sheep	human clotting factors produced	treatment for haemophilia
goat	human antithrombin produced	treatment for thrombosis
chickens	monoclonal antibodies produced	vaccine production
bacteria	human insulin produced	treatment for diabetes
bacteria	human growth hormone produced	treatment for dwarfism
soybean	resistance to herbicide	increased yield
tobacco	resistance to tobacco mosaic virus	increased yield
potato	pesticide produced	increased yield

 Look up www.pbs.org/wgbh/harvest/engineer/transgen.html

LET'S THINK ABOUT THIS

The information below outlines some of the stages that genetic engineers would use to transfer a useful gene from a bacterial cell to a crop plant.

Stage	Description
A	Plant cells containing gene X are grown into plantlets.
B	The position of gene X is located.
C	Gene X is transferred into the nucleus of the plant cell.
D	Chromosome containing gene X is extracted from a bacterial cell.
E	Gene X is cut out from the chromosome.

(a) Put the stages into the correct order.

(b) Which enzyme would be used to cut gene X out of the bacterial cell?

(c) How could the genetic engineers locate gene X on the chromosome?

(d) What are the benefits of using genetic engineering rather than selective breeding?

DON'T FORGET

If genetic engineering is to be successful, it is critical that useful genes can be located easily. You must be aware of the use of (i) **gene probes** and (ii) **recognition of banding pattern** to locate genes on chromosomes.

LET'S THINK ABOUT THIS

Genetic engineering is not just of use in treating human diseases or for improving the quality and quantity of agricultural produce. New materials are being developed using genetically-engineered spider silk. A spider silk gene has been inserted into a goat's egg before fertilisation, with the result that goat's milk containing silk fibre proteins has been produced. The silk fibres can be used to make lightweight bullet-proof vests.

MAINTAINING A WATER BALANCE – ANIMALS

ADAPTATION

adaptation
a feature of an organism that aids survival in its environment

structural	**physiological**	**behavioural**
specialised structures	the way the body works	response to environmental stimuli

OSMOREGULATION

Osmoregulation is the maintenance of water balance in an organism.

- When **hypotonic** to the environment, an organism loses water by osmosis.
- When **hypertonic** to the environment, an organism gains water by osmosis.

Comparison of osmoregulation in fresh-water and salt-water fish

Feature	Type of fish	
	Fresh water	**Salt water**
glomeruli	large number of large glomeruli	low number of small glomeruli
filtration rate	high	low
filtrate	dilute	concentrated
chloride secretory cells	actively absorb salts from the water	actively excrete salts into the water

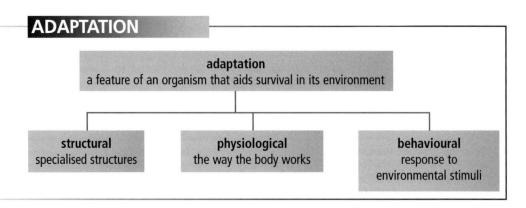

salts absorbed through active transport by chloride secretory cells in gills

water moved to gills / kidney

water enters by osmosis

dilute urine, large volume

large glomerulus gives high filtration rate

salts excreted through active transport by chloride secretory cells in gills

water moved to gills and some drunk / kidney

water leaves by osmosis

concentrated urine, low volume

small glomerulus gives low filtration rate

> **DON'T FORGET**
>
> Fresh-water fish are hypertonic to their environment and gain water constantly by osmosis.
> Salt-water fish are hypotonic to their environment and lose water constantly by osmosis.

contd

OSMOREGULATION contd

Water balance in migratory fish

Salmon and eels are examples of fish that can efficiently regulate their water balance in both fresh- and salt-water environments. During a salmon's life cycle, it moves from fresh water to salt water, and back again. During the transition from fresh water to salt water, the rate of glomerular filtration decreases, and there is a decrease in the rate of salt uptake by chloride secretory cells. This effect is reversed when the fish returns to fresh water again.

WATER CONSERVATION IN A DESERT MAMMAL

The kangaroo rat is an example of a desert mammal that demonstrates efficient water conservation.

Type of adaptation	Form of adaptation	Reasons for adaptation
physiological	dry mouth and nasal passages no sweat	reduce water loss
physiological	long loops of Henlé high level of antidiuretic hormone increased reabsorption of water from waste in large intestine dry mouth inability to sweat	increase water retention
behavioural	underground burrow	reduce water loss
behavioural	active at night and inactive during the day	keep cool

DON'T FORGET

Water loss must equal water gain.

LET'S THINK ABOUT THIS

1. Complete the following table, which shows adaptations for maintaining water balance in two species.

Organism	Why adaptation is required	Kidney adaptation	Urine production
desert mammal	animal loses water to the environment		low volume
fresh-water fish		many, large glomeruli	

2. The list below shows some adaptations of bony fish for osmoregulation.

   A   low filtration rate in kidney

   B   high filtration rate in kidney

   C   few, small glomeruli

   D   many, large glomeruli

   E   active uptake of salts by chloride secretory cells

   F   active secretion of salts by chloride secretory cells

   Which adaptations would be found in salt-water fish?

MAINTAINING A WATER BALANCE – PLANTS

TRANSPIRATION STREAM

Transpiration is the continuous movement of water from the roots to the leaves and out of the plant through the stomata. The rate of transpiration can be affected by temperature, wind, relative humidity and atmospheric pressure. It relies upon a number of forces and assists in the movement of ions throughout the plant.

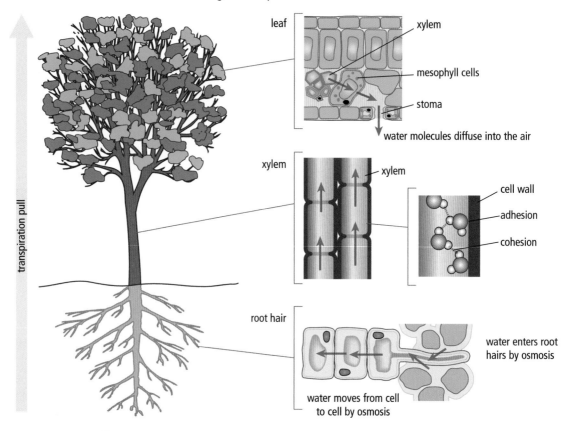

leaf
xylem
mesophyll cells
stoma
water molecules diffuse into the air

xylem
xylem
cell wall
adhesion
cohesion

root hair
water enters root hairs by osmosis
water moves from cell to cell by osmosis

transpiration pull

Forces

The movement of water up the xylem is brought about by **transpiration pull** and is reliant on two forces:

- **adhesion** – water molecules stick to the xylem
- **cohesion** – water molecules bind to each other.

Transpiration pull is caused by a low water concentration in the leaf cells pulling water up the plant stem through the xylem vessels.

DON'T FORGET

Transpiration pull is similar in action to drinking from a straw.

Look up www.phschool.com/science/biology_place/labbench/lab9/xylem.html
www.kscience.co.uk/animations/transpiration.swf

STOMATAL MECHANISM

Through the stomatal mechanism, the plant demonstrates a limited degree of **osmoregulation**.

State of guard cells		Stomata	Light level
turgid	water has moved into guard cell	open	light
flaccid	Water has moved out of guard cell	closed	darkness

ADAPTATION IN XEROPHYTES AND HYDROPHYTES

Xerophytes					
Environmental condition – hot, dry, arid conditions or exposed, windy conditions					
Effect	decreased evaporation		retention of water	increased water uptake	drought survival
Adaptation	thick epidermis waxy leaves few leaves needle leaves few stomata	hairy leaves rolled leaves sunken stomata reversed stomatal rhythm	succulent tissue	superficial roots long tap roots	dormancy

Hydrophytes		
Environmental condition – partially or completely submerged in water		
Effect	gas exchange	stem kept upright	reduced damage by water current
Adaptation	long leaf stalks stomata on upper surface no cuticle on submerged leaf	air spaces in stem	xylem reduced and located in centre submerged leaves

LET'S THINK ABOUT THIS

Water evaporation from the leaf helps to keep the plant cool – another useful function of transpiration. However, if more water is lost through the stomata than is being absorbed through the roots, leaf cells lose turgor and the stomata close. So, stomata can be closed even on a hot, sunny day.

OBTAINING FOOD – ANIMALS

Animals are **heterotrophic**, as they have to forage for food. They therefore have to be **mobile**.

Plants remain fixed in one position and are therefore immobile (**sessile**), and are **autotrophic**, as they are able to make their own food during photosynthesis.

FORAGING BEHAVIOUR

Foraging is the act of obtaining food by an animal. Before it eats, the animal has problems to overcome. It has to decide **where** its food might be found, **when** food may be available and **what** the food might be. For foraging to be effective, the energy gained must be greater than the energy used in obtaining food. The table below shows the foraging behaviour of some species.

Organism	Behaviour pattern when foraging
Cattle	To balance their need for water and food, cattle will forage within a circle with a water source at its centre. This allows them to look for an energy efficient source of food without running out of water.
Orb-web spiders	These spiders locate their web where insects are likely to fly, and sit and wait on food coming to them. The spider assesses the capture rate and if the location produces too low a yield of insects, the spider moves the web to try somewhere else.
Blackbirds	A blackbird takes short 'runs' on either alternate legs or by hopping. It then stops and examines the ground looking for prey, which it picks up with its bill. If the bird is successful, it takes shorter 'runs'.
Tiger beetle larvae	This beetle sits in a burrow which it blocks with the top of its head. When prey comes close, the beetle grabs it with powerful jaws, minimising the energy expenditure.

Foraging can be affected by the richness of the food supply, choice of prey size and the risk of predation.

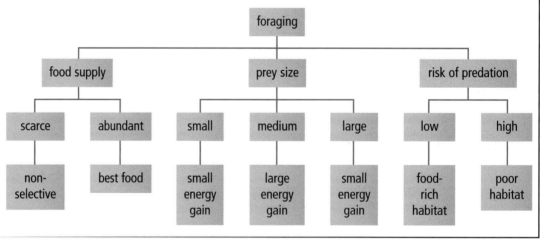

INTERSPECIFIC AND INTRASPECIFIC COMPETITION

When animals forage for the same food, they are said to be in **competition** with each other for that food. If the foraging animals are of the same species, for example two foxes hunting the same rabbits, this is called **intraspecific** competition. If the two animals are of different species, as in the case of a fox and a hawk hunting the same rabbits, this is called **interspecific** competition. Intraspecific competition is more competitive than interspecific competition because organisms of the same species require exactly the same resources whereas different species have fewer resources in common.

The competition exclusion principle

Interspecific competition for a particular resource can result in one species being so unsuccessful in obtaining food that it is forced to migrate out of the area or face extinction.

DON'T FORGET

- Intraspecific – same species compete.
- Interspecific – different species compete.

DOMINANCE HIERARCHY AND COOPERATIVE HUNTING

Intraspecific competition can be deliberately reduced by some species, allowing all individuals in a group to obtain the resources that they require. Large social groups of animals often manage to coexist by forming a **dominance hierarchy**. Here the group is organised by the level of dominance of individuals, with the most aggressive individual being most dominant and the least aggressive individual being the most subordinate. For example, a pecking order is established in chickens by determining who pecks whom. Every bird remembers its dominance status, knowing whom it can dominate and whom it must defer to.

In **cooperative hunting**, animals work together to obtain food. Where a dominance hierarchy exists, subordinate animals may gain more food than by hunting alone using this method. For example, hyenas use an ambush strategy hiding in cover ready to pounce as the rest of the group drive the prey towards them. Sandtiger sharks also use cooperative hunting, working together to surround schools of fish and herding them to shallow water where it is easier to catch them.

Dominance hierarchies and cooperative hunting allow:

- all individuals in the group to obtain food, which might not occur if foraging alone.

- the possibility of injury, through fighting for food and other resources, to be reduced.

- the best adapted individuals to breed and pass on their genes.

TERRITORIAL BEHAVIOUR

A **territory** is an area held and defended by an animal or group of animals against animals of the same species. It means that each mating pair of animals and their offspring are adequately spaced to receive a share of available resources.

Animals can defend their territories through:

- Social signals

- Fighting

- Marking

For example: A gull defends its territory (nest) by fighting with other gulls.

DON'T FORGET

Territorial behaviour is a feature of intraspecific competition. It is often solitary whereas a dominance hierarchy involves animals living together.

 LET'S THINK ABOUT THIS

Planaria Investigation

You may have carried out an investigation to study the foraging behaviour of planaria. You should be able to explain the random movement of planaria when food is absent, and their direct movement to a food source by use of chemoreceptors.

OBTAINING FOOD – PLANTS

SESSILITY IN PLANTS; MOBILITY IN ANIMALS

Animals are **heterotrophic**, as they are mobile and have to forage for food. Plants remain fixed in one position (**sessile**) and are **autotrophic**, as they are able to make their own food during photosynthesis.

Plants	Animals
produce their own food	can forage for food
sessile	mobile
cannot shelter from adverse conditions	can shelter from adverse conditions
require light	do not depend on light

COMPETITION IN PLANTS

Plants compete for four main resources when growing in the same habitat:

- light
- water
- soil minerals
- space.

Intraspecific competition between densely-populated plants is intense, as plants of the same species require identical resources. Overlapping of leaves results in individual plants receiving less light, while competition for water and minerals increases due to interweaving of root systems.

Interspecific competition is less intense, as different plant species require different nutrients. Interspecific competition can result in adaptation of plants. Two such adaptations are:

- diversification of species to require different nutrients
- production of toxic substances that inhibit the growth of other species.

Look up
http://www.agron.iastate
.edu/plantscience/plant_
competition_ photos.htm

THE EFFECT OF GRAZING ON SPECIES DIVERSITY

Consider three levels of grazing in an ecosystem.

low grazing levels —— When few plants are being damaged by grazing, interspecific competition for resources allows the best-adapted plants to dominate and displace other species. As a result, the overall number of species will be low.

medium grazing levels —— The dominant plants are eaten, allowing less well-adapted plants more access to resources such as light and space. The diversity of species will increase.

high grazing levels —— Some plant species are unable to tolerate the damage inflicted by large numbers of grazing animals. The diversity of species decreases.

Within a pasture, grasses can tolerate grazing because shoots regrow quickly from ground level. Diversity of species is maintained due largely to:

- the unpalatability of some species
- some plant species growing so close to the ground that they avoid being eaten.

COMPARISON OF COMPENSATION POINT IN SUN AND SHADE PLANTS

In plants, the light intensity at which the rate of photosynthesis is equal to the rate of respiration is called the **compensation point**. At this point, there is neither a net gain nor a net loss of carbon dioxide, and all food produced during photosynthesis is used up in respiration. A plant's respiration rate remains relatively constant, but the rate of photosynthesis is dependent on light intensity and so varies throughout the day. For most plants, the compensation point is reached during early morning and then again in late evening.

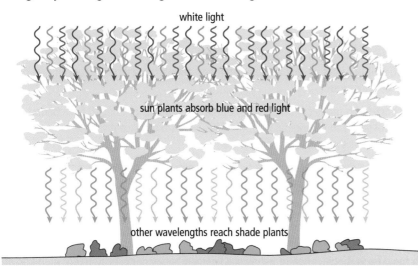

white light

sun plants absorb blue and red light

other wavelengths reach shade plants

The light intensity at which sun and shade plants reach their compensation point is different. Shade plants reach their compensation point at lower light intensities, and therefore earlier in the day than the sun plants. This gives the shade plants more time to produce and store food before the light intensity decreases again.

Some woodland shade plants produce flowers in early spring because their rate of photosynthesis is higher when the trees above have not yet produced leaves. So, more light is able to reach the woodland floor.

Shade plants are also more efficient at using green light for photosynthesis than sun plants. This is important, as red and blue light are absorbed by the sun plants in the canopy layer, leaving the transmitted green light to reach the lower plant layers.

DON'T FORGET

Sun plants are those plants that thrive in light conditions.
Shade plants are those plants that thrive in darker, shaded conditions.

LET'S THINK ABOUT THIS

You should be able to describe an experiment to investigate the response of leaf discs from sun and shade plants on exposure to green light.

Leaf discs are placed in a syringe containing sodium hydrogen carbonate solution. When the air is removed from the syringe, the leaf discs sink. Exposure to white light allows photosynthesis to occur in both sun and shade plants, resulting in oxygen being produced, and the leaf discs float. However, if green light is used, only the shade plants produce oxygen by photosynthesis and float. Sun plants remain at the bottom of the syringe.

Use your experimental notes to explain each step in this investigation.

Plants arrange their leaves in a mosaic pattern. Why? Because this pattern increases the surface area exposed to light and reduces overshadowing.

COPING WITH DANGERS – ANIMALS

AVOIDANCE BEHAVIOUR AND HABITUATION

Avoidance behaviour is a behavioural adaptation which reduces the risk of an organism being eaten. This is demonstrated by the marine snail *Aplysia*, which withdraws its gill when it is touched.

If this stimulus is repeated frequently, the response is suppressed and the snail no longer withdraws its gill. This modification of behaviour is known as **habituation**. Habituation is the process by which an animal becomes accustomed to a non-threatening stimulus and ignores it. This is important, as it prevents energy being wasted and allows the animal to concentrate on stimuli that may be dangerous.

Habituation is **short-term** modification of behaviour: if the stimulus is stopped for a period of time, the animal will revert to its original response. This allows a defence response to occur when there is more chance that the stimulus is dangerous.

Habituation can also be demonstrated in territorial birds. When repeatedly played the birdsong of a potential challenger, birds' responses decline over a period of time.

LEARNING IN HUMANS

Learning is a **long-term** modification of behaviour. Learned behaviour is acquired or modified as a result of an individual's experiences during its lifetime. Human learning patterns can be divided into three groups.

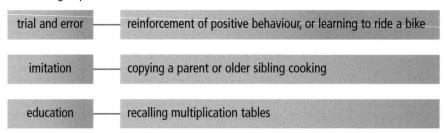

trial and error	reinforcement of positive behaviour, or learning to ride a bike
imitation	copying a parent or older sibling cooking
education	recalling multiplication tables

Learning curves

Learning curves are used to study the effect of practice on the acquisition of motor skills. Here, learning is measured by either:

- measuring the time taken to complete the task, or
- counting the number of errors made in completing the task.

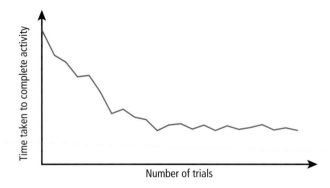

The graph shows that, as the number of trials increases (and the individual practises the task), performance improves until the task cannot be carried out any faster.

INDIVIDUAL AND SOCIAL MECHANISMS FOR DEFENCE

Individual mechanisms for defence

Method of defence	Animal	Method	Success
active	horse	runs at fast speed	outruns predators
	squid	produces an inky fluid	confuses predators and escapes
	deer	uses antlers to charge at competitor or predator	wound or scare opponent
	some snakes	poisonous venom	inject poison
	skunk	putrid-smelling fluid produced from scent glands	causes distress to predators
passive	porcupines	protective covering of spines	discourages attack by predator
	moths	markings mimic menacing eyes	looks like owl eyes to scare off predator
	trout	counter shading – two-colour tones	prevents animals being seen
	tree frog	bright green colour acts as camouflage	animal is difficult to see against green leaves
	caterpillars	irritating hairs	predators find it difficult to eat the caterpillar

Social mechanisms for defence

'Safety in numbers' helps to protect animals because:

- the group can be organised so that a lookout is always available to provide an alarm call

- it is more difficult for predators to identify and pick off individuals if they are part of a group

- a group can work together to mob or attack a predator and protect the young.

Some examples of these defence mechanisms are given below.

birds	a flock moves as a single organism, making it difficult for a predator to pick out an individual
elephants	form a protective group to defend the herd – adults turn with tusks outwards to protect the calves in the centre of the group
meerkats	individuals act as lookouts to warn of approaching danger; when challenged by a rival gang, meerkats demonstrate mob behaviour, grouping together and jumping up and down to scare away rivals
chimpanzees	form a mob attack on neighbouring groups, with extreme brutality, throwing rocks, charging at the rivals and fighting

LET'S THINK ABOUT THIS

Try to write bullet points on methods of defence under the following headings:

1. Passive methods – (i) surface markings (ii) protective coverings (iii) chemicals

2. Active methods – (i) speed (ii) distraction (iii) chemical attack (iv) feigning death

3. Social methods

COPING WITH DANGERS – PLANTS

STRUCTURAL DEFENCE MECHANISMS

Plants are sessile and are, therefore, unable to move away when they are attacked by animals. Plants have adapted to produce structural features that are designed to discourage grazing.

Stings

Example: nettles

Mechanism: when the stinging hair is touched, the tip breaks off, converting the hair into a needle that penetrates the skin and injects an irritant

Thorns

Examples: rose bushes, gooseberry, hawthorn

Mechanism: the thorns damage the inside of the animal's mouth, discouraging further attacks

Spines

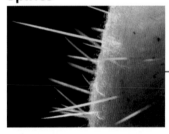

Examples: holly, cacti, gorse

Mechanism: spines form a barrier that prevents grazing animals from reaching delicate tissues

HOLLY INVESTIGATION

What is the relationship between the number of spines on a holly leaf and its height above ground?

1. Branches are collected from different heights of an unpruned holly tree.

2. The number of spines around the margin of each leaf is counted. An average number of spines per leaf is calculated for each branch height.

Conclusion – as the height of the branch increases, the number of spines decreases.

Explanation – holly trees with many spines on their lower leaves are less likely to be damaged by grazing animals. The higher branches are attacked less frequently and so require less protection.

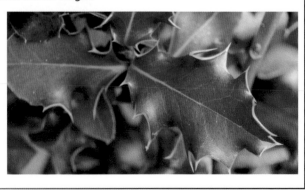

ABILITY TO TOLERATE GRAZING

For a plant to recover successfully from grazing, leaves must regrow rapidly so that photosynthesis can start again. Plants have evolved several mechanisms that promote rapid regrowth.

low meristems	Plants such as grasses have many buds or meristems near the base of the plant, allowing large numbers of shoots to rapidly grow out after grazing damage.
deep roots	Deep roots remain undamaged by grazing and allow rapid regeneration of the plant.
rhizomes	Plants such as knotweed and some grasses grow horizontal underground stems called rhizomes that act as storage organs. Rhizomes can produce the shoot and root systems of new plants and allow rapid spread.

LET'S THINK ABOUT THIS

1. Explain why natural selection in holly trees would favour plants with many spines on the leaves of lower branches.

2. Explain how stinging nettles sting.

3. How does a hawthorn tree deter grazers?

4. What is the role of a rhizome?

5. Why do low meristems help plants to tolerate grazing?

LET'S THINK ABOUT THIS

Some plants avoid damage by growing low to the ground. Dandelions grow in a rosette pattern, with leaves on very short stems. As they grow, the leaves push down where they are less likely to draw the attention of grazing animals.

GROWTH DIFFERENCES BETWEEN PLANTS AND ANIMALS

Growth of an organism involves an increase in dry mass through:

1. an increase in cell numbers

2. an increase in cell size.

growth and regeneration in animals	<ul><li>Mitosis occurs in most tissues, allowing growth over the whole body.</li><li>Growth continues until adulthood.</li><li>Mammals have limited powers of regeneration.</li><li>Mammals do not have meristems.</li></ul>
growth and regeneration in plants	<ul><li>Growth occurs at points called **meristems**.</li><li>A meristem is a cluster of undifferentiated cells that can divide by mitosis.</li><li>Plants continue to grow throughout life.</li><li>Plants have extensive powers of regeneration.</li></ul>

PLANT MERISTEMS

There are two types of meristem:

1. apical meristems – located at the tips of roots and shoots

2. lateral meristems – containing undifferentiated cells called **cambium**, and located between the xylem and phloem in vascular bundles.

Apical meristems

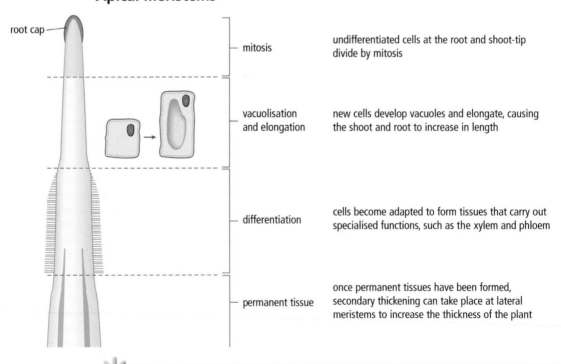

root cap

mitosis — undifferentiated cells at the root and shoot-tip divide by mitosis

vacuolisation and elongation — new cells develop vacuoles and elongate, causing the shoot and root to increase in length

differentiation — cells become adapted to form tissues that carry out specialised functions, such as the xylem and phloem

permanent tissue — once permanent tissues have been formed, secondary thickening can take place at lateral meristems to increase the thickness of the plant

Look up www.wadsworthmedia.com/biology/0495119814_starr/big_picture/ch25_bp.swf

contd

PLANT MERISTEMS contd

Lateral meristems

Lateral meristems are present as **cambium** cells, lying between the xylem and phloem in **vascular bundles**. Here, clusters of undifferentiated cells give rise to new xylem and phloem cells, increasing the width of the stem (**secondary thickening**).

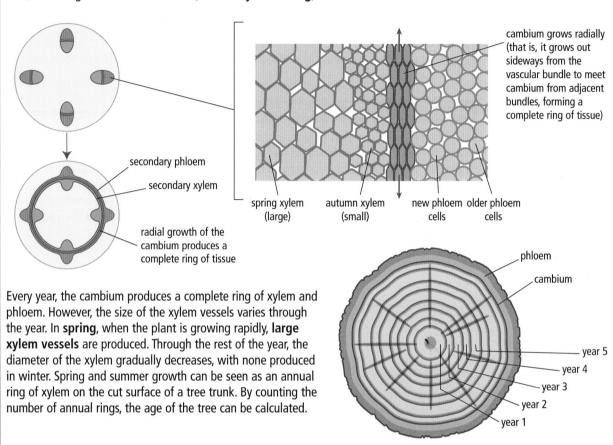

cambium grows radially (that is, it grows out sideways from the vascular bundle to meet cambium from adjacent bundles, forming a complete ring of tissue)

secondary phloem

secondary xylem

radial growth of the cambium produces a complete ring of tissue

spring xylem (large)

autumn xylem (small)

new phloem cells

older phloem cells

Every year, the cambium produces a complete ring of xylem and phloem. However, the size of the xylem vessels varies through the year. In **spring**, when the plant is growing rapidly, **large xylem vessels** are produced. Through the rest of the year, the diameter of the xylem gradually decreases, with none produced in winter. Spring and summer growth can be seen as an annual ring of xylem on the cut surface of a tree trunk. By counting the number of annual rings, the age of the tree can be calculated.

phloem

cambium

year 5

year 4

year 3

year 2

year 1

LET'S THINK ABOUT THIS

You must be able to identify and explain the growth patterns of an annual plant, a tree, a human and a locust.

Annual plant

Dry mass / Time

plant loses mass as energy stores are used for growth until leaves develop

loss of flowers and leaves cause decrease in mass

photosynthesis takes place and mass increases

Tree

Height / Time

The height of a tree increases throughout its life. Most yearly growth occurs during spring and summer (most rapid photosynthesis), and in autumn the mass decreases if leaves are shed.

Humans

Height / Time

girls

boys

Although humans grow until adulthood, there are two periods of rapid growth (**growth spurts**): from birth to year two, and during puberty.

Locust

Length / Time

new exoskeleton prevents further growth

exoskeleton is shed, allowing an increase in body growth

GENETIC CONTROL

You should remember that proteins are coded for by the sequence of bases (**genes**) on a molecule of DNA (see page 26). As all cells inherit complete sets of genetic information during mitosis, each cell has the potential to produce protein from every gene in the code.

However, different types of cell only make proteins required by that cell type for normal function. This means there must be a mechanism that allows **genes** to be '**switched on**' and '**switched off**'.

JACOB–MONOD HYPOTHESIS

Two scientists, Jacob and Monod, first suggested a hypothesis to explain how genes 'switch on and off'. The organism they studied was the bacterium *Escherichia coli* (**E. coli**). *E. coli* synthesises an enzyme called β-**galactosidase**, which catalyses the breakdown of the milk sugar **lactose** into **glucose** and **galactose**. Jacob and Monod found that β-galactosidase is only produced when a bacterium detects that lactose is present, preventing unnecessary use of resources and energy by the cell. So, how does a bacterium control expression of the gene for β-galactosidase?

The diagram below shows the genes that are involved.

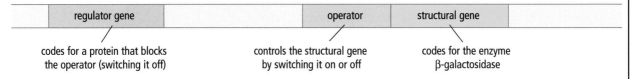

regulator gene		operator	structural gene
codes for a protein that blocks the operator (switching it off)		controls the structural gene by switching it on or off	codes for the enzyme β-galactosidase

Lactose absent

When lactose is absent, a repressor molecule (coded for by the regulator gene) binds with the operator. Because the operator gene is switched off, the structural gene remains switched off, preventing transcription of the genetic code for the enzyme β-galactosidase, and so the enzyme is not produced.

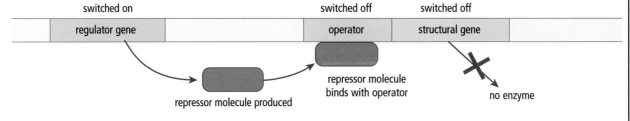

Lactose present

When lactose is present, the repressor molecule binds to lactose instead of the operator. The operator now switches on the structural gene, allowing transcription of the code for β-galactosidase, and the enzyme is manufactured.

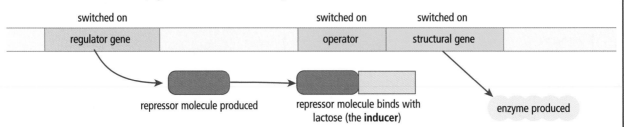

When all the lactose has been broken down, the repressor molecule binds to the operator gene, switching it off again.

contd

Look up www.sumanasinc.com/webcontent/animations/content/lacoperon.html

LET'S THINK ABOUT THIS

You should understand how expression of the β-galactosidase gene is controlled.

1. Identify which of the following statements is true.
   (a) The regulator gene codes for β-galactosidase.
   (b) The repressor binds to the inducer.
   (c) The structural gene codes for the inducer.
   (d) The operator binds to the repressor.
   (e) β-galactosidase breaks down the inducer.
   (f) The structural gene is switched on in the presence of the inducer.

2. Explain how lactose acts as an inducer for the enzyme β-galactosidase.

LET'S THINK ABOUT THIS

The effects of β-galactosidase on milk can be studied using immobilised enzyme techniques.

1. A syringe is filled with a mixture of sodium alginate and the enzyme β- galactosidase.

2. Droplets of the mixture are slowly released into a calcium chloride solution, where they solidify into beads.

3. The alginate beads are then rinsed in distilled water.

4. The beads are packed into a column.

5. Milk is slowly dripped through the column and collected.

6. The treated milk is then tested for the presence of glucose using a Clinistix test strip. A positive test is indicated by a colour change to purple–blue within 10 seconds.

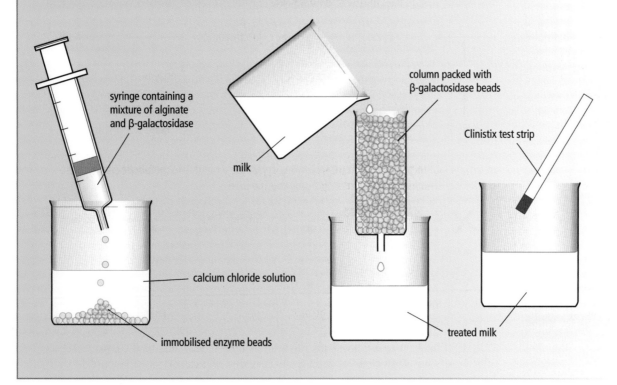

CONTROL OF METABOLIC PATHWAYS AND CELL DIFFERENTIATION

CONTROL OF METABOLIC PATHWAYS

All the chemical reactions that take place in an organism make up the **metabolism**.

A **metabolic pathway** is a series of chemical reactions that follow on, one after another. Each stage in the pathway is controlled by an enzyme, with the product of one reaction becoming the substrate for the next reaction.

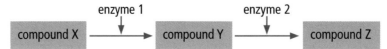

In the diagram above, compound Z can only be produced if both enzyme 1 and enzyme 2 are present.

Each enzyme is coded by a different gene. When the base sequence of one gene is faulty (see gene mutations, page 40), the enzyme is not produced and the metabolic pathway will be blocked.

This is called an **inborn error of metabolism**. Inborn errors of metabolism occur when the individual is homozygous for the faulty gene.

In the above example, if enzyme 2 is not present, compound Y builds up and compound Z is not produced.

Phenylketonuria (PKU)

PKU is an example of an inborn error of metabolism, where absence of a specific enzyme in a metabolic pathway results in failure to break down the amino acid **phenylalanine** (taken in through the diet) and its accumulation in the body. Sufferers inherit two copies of a mutated gene.

The normal metabolic pathway is shown below.

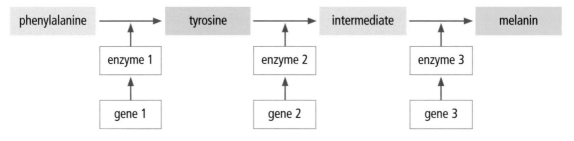

In sufferers of PKU, gene 1 is mutated and enzyme 1 is absent. The metabolic pathway is blocked, preventing production of tyrosine and melanin.

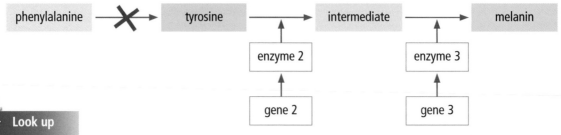

Look up
www.ygyh.org/pku/cause.htm

With the normal pathway blocked, phenylalanine is broken down into toxic substances that affect brain cells, causing mental disability and seizures. PKU can be controlled by eating a diet low in phenylalanine.

CONTROL OF CELL DIFFERENTIATION

After fertilisation, the resulting zygote contains all the genetic information required to make every cell type in the organism. As mitosis proceeds to produce a multicellular organism, differentiation produces different cell types that are capable of carrying out specific functions. Some proteins are made in every cell, and the genes that code for these proteins are switched on irrespective of cell type, for example the genes for proteins involved in respiration. Other proteins are made only in specific cell types, meaning that these genes are switched on in the cell types that make the protein and are switched off in all other cells.

For example, as mammalian development progresses, some cells become differentiated and switch on the MyoD gene. MyoD protein is required in muscle cells.

LET'S THINK ABOUT THIS

1. Explain what is meant by the term 'inborn error of metabolism'.

2. Build-up of phenylalanine in babies with PKU does not occur until after birth.

   (i) Explain why there is no build-up of phenylalanine before birth.

   (ii) Explain why phenylalanine builds up after birth.

3. What is the importance of 'switching on or off' genes during cell differentiation?

LET'S THINK ABOUT THIS

Sufferers of PKU have very fair skin and blue eyes, but are not albinos. Why?

Although phenylalanine cannot be broken down in PKU, some tyrosine forms part of the normal diet. This tyrosine can be broken down to produce melanin, which gives some pigmentation to the skin.

HORMONAL INFLUENCES 1

HORMONAL CONTROL IN ANIMALS

Hormones

Endocrine glands produce **hormones** (chemical messengers) that are released into the **bloodstream** and transported to the **target cells** on which they act. Any tissue containing cells with the proper receptors for a particular hormone will be affected. The effect of a hormone continues until it is broken down by the liver.

Pituitary gland and pituitary hormones

The **pituitary gland** lies at the base of the brain, protected in its own little bony hollow in the skull. This gland is extremely important because it produces many hormones, including two that are particularly important for growth and development: **growth hormone (GH)** and **thyroid-stimulating hormone (TSH)**.

Growth hormone (GH) acts on bone cells and soft-tissue cells, **increasing the rate of amino acid uptake and protein synthesis**. As a result, proteins are synthesised rapidly, promoting growth of **muscle** and **bone** tissues.

The table below shows the effects of abnormal levels of growth hormone on bone growth.

Level of production	Effect on bone growth
under-production of growth hormone during adolescence	overall growth is reduced – individuals have small body size but normal body proportions
over-production of growth hormone during adolescence	overall growth is increased – individuals have a large body size but normal body proportions
over-production of growth hormone after adolescence	hand, foot and jaw bones are noticeably enlarged

Thyroid-stimulating hormone (TSH) acts on cells in the **thyroid gland**. The thyroid gland responds by producing other hormones, including **thyroxine**. Thyroxine increases the **metabolic rate** of body cells, which stimulates growth.

Under-secretion of TSH causes a decrease in thyroxine production and a drop in the metabolic rate. In children, severe under-secretion can cause a form of dwarfism which is characterised by stunted bone growth and mental disability.

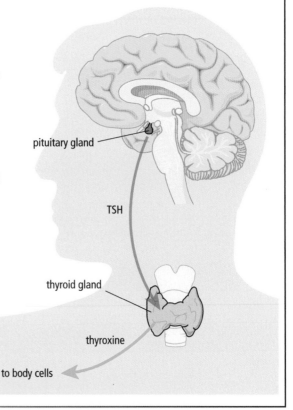

pituitary gland

TSH

thyroid gland

thyroxine

to body cells

HORMONAL CONTROL IN PLANTS

Unlike animals, plants do not have specialised organs for hormone production. Instead, plant hormones are made in the leaves, root and stem tips, flowers, fruits and seeds.

Gibberellic acid

Gibberellic acid (GA) is a plant hormone that has roles in a seed's emergence from a dormant state and on plant growth.

DON'T FORGET

GA does not increase the number of internodes.

effect on dwarf varieties

The presence of GA results in elongation of the internodes (the section of stem between lateral buds or leaves), causing the height of the plant to match that of normal varieties.

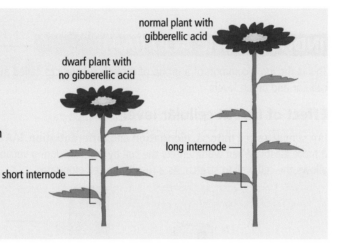

effect on dormancy

When conditions are favourable for germination, GA brings a seed out of the dormant state. The embryo produces GA and releases it into the **aleurone layer** of the seed. The aleurone layer is stimulated to produce α-**amylase**, which catalyses the breakdown of stored starch into maltose, used by the embryo as a source of energy in respiration.

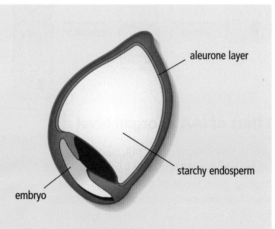

LET'S THINK ABOUT THIS

1. Which glands produce (i) thyroxine (ii) growth hormone (iii) TSH?

2. On which tissues does growth hormone exert the greatest effect?

3. What is the role of thyroxine in the body?

4. Describe how α-amylase production is induced in a seed.

5. Explain why α-amylase production is important in the process of germination.

LET'S THINK ABOUT THIS

The level of thyroxine in the blood is controlled by a negative feedback loop. When the level gets too high, thyroxine inhibits production of TSH by the pituitary gland. This, in turn, causes a decrease in thyroxine production until normal levels of thyroxine are restored.

HORMONAL INFLUENCES 2

Plants do not have specialised organs for hormone production; plant hormones are made, in very small concentrations, in the leaves, root and stem tips, flowers, fruits and seeds. The effect of one hormone is dependent not only on its concentration but also on its relative concentration compared to other hormones.

INDOLE ACETIC ACID (IAA)

This is the most common of a group of growth substances called **auxins**. IAA can act at both cellular and organ levels.

Effect of IAA at cellular level

IAA stimulates cell **mitosis**, **elongation** and **differentiation**. IAA acts on the cell wall, making it more elastic. When water enters the cell by osmosis during vacuole formation, this elasticity allows the cell wall to stretch. As a result, the cell elongates.

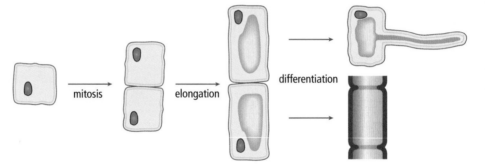

mitosis   elongation   differentiation

Effect of IAA at organ level

apical dominance

Apical meristems produce IAA, causing the apical bud to lengthen. High levels of IAA moving down the plant from the apical bud inhibit development of lateral buds. Therefore, the plant grows upwards from the apical bud, rather than producing the lateral branches that would make it bush outwards. The plant displays apical dominance.

If the apical bud is removed, the level of IAA drops and inhibition of lateral buds decreases. So, the plant grows lateral branches. If IAA is applied to the site of the severed apical bud, growth of the lateral buds is inhibited.

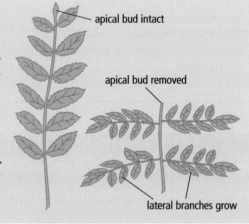

apical bud intact

apical bud removed

lateral branches grow

leaf abscission

Leaves stay attached to the plant while IAA is present. When the level of IAA decreases, a layer of cells at the base of the leaf stalk forms an abscission layer. Here, the cell walls become weakened and, eventually, the leaf falls off.

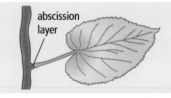

abscission layer

fruit formation

IAA is produced by seeds after fertilisation. IAA stimulates development of the ovary wall to produce the fruit.

Look up http://library.thinkquest.org/C006669/data/Biol/phormone_1.html

INVESTIGATING THE EFFECT OF IAA ON ROOT LENGTH IN MUSTARD SEEDLINGS

This experiment can be used to identify the effect of different IAA concentrations on root length.

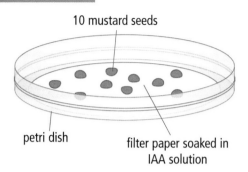

10 mustard seeds

petri dish

filter paper soaked in IAA solution

1. A serial dilution of IAA is made, with concentrations ranging from 10^{-4} parts per million (ppm) to 1 ppm. Note that IAA solution is light-sensitive and must be stored in the dark to prevent loss of activity.

2. Ten mustard seeds are placed in petri dishes containing different IAA concentrations. Using ten seeds at each concentration increases reliability. Different syringes must be used for each concentration to prevent contamination of solutions.

3. A control experiment is set up using distilled water instead of IAA solution.

4. The seeds are allowed to grow for three days in the dark at 30°C.

5. Root length is measured by laying out each seedling on an acetate of graph paper.

6. Where the average root length is greater than that of the control group, there has been stimulation. If average root length is less than in the control group, IAA has inhibited elongation of roots.

7. To calculate the percentage stimulation or inhibition:

$$\frac{\text{difference between experimental and control average values}}{\text{average length of control roots}} \times 100$$

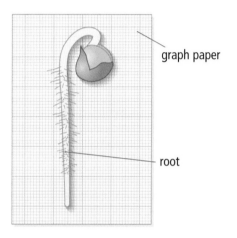

graph paper

root

COMMERCIAL USES OF PLANT HORMONES

Auxins can be used in several ways to improve the yield or quality.

rooting powder	Roots can be induced to grow from the cut ends of stems if rooting powder containing auxins is applied. This is used when taking cuttings.
herbicides	When sprayed on plants, auxins can stimulate plant growth so much that the plant uses up its food supply and dies. Selective weedkillers for lawns contain auxins, as the broad-leaved weeds absorb more of the auxin than the grass.
fruit development	When treated artificially with auxins, unfertilised flowers develop into seedless fruit.
delayed abscission	Auxins can be sprayed onto fruit crops to prevent the abscission layer from developing.

LET'S THINK ABOUT THIS

When a shoot is exposed to light from directly above, it grows straight upwards. However, if the light comes from one side, the shoot bends towards the light. This is called **phototropism** and results from the accumulation of auxins on the shaded side of the shoot. The higher concentration of auxins here causes the cells of the shaded side to elongate more than the cells on the exposed side, and the shoot bends.

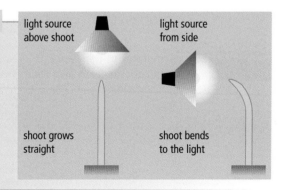

light source above shoot

light source from side

shoot grows straight

shoot bends to the light

ENVIRONMENTAL INFLUENCES ON PLANTS

MINERAL DEFICIENCIES IN PLANTS

Macro-elements are substances that are required by plants for normal growth. These include:

- **nitrogen**
- **magnesium**
- **phosphorus**
- **potassium.**

The role of these macro-elements in normal plant metabolism and the symptoms of deficiency are shown below.

nitrogen

Role in metabolism
Nitrogen is needed for the synthesis of amino acids, proteins, ATP and nucleic acids, such as DNA.

Symptoms of deficiency
Lack of nitrogen leads to stunted plant growth and elongation of the roots to form a long, thin root system. Older leaves show the deficiency first, as nitrogen is recycled in the plant to make new proteins in young leaves. The leaves turn yellow (**chlorosis**), with a red leaf base.

magnesium

Role in metabolism
Magnesium is the central atom in the chlorophyll molecule, so is required for the formation of chlorophyll.

Symptoms of deficiency
As chlorophyll is essential for photosynthesis, a deficiency in magnesium causes stunted plant growth. Older leaves show the deficiency first, with chlorosis occurring between the leaf veins, eventually spreading to cover the entire leaf.

phosphorus

Role in metabolism
Phosphorus is used to synthesise nucleic acids and ATP.

Symptoms of deficiency
Deficiency causes stunted plant growth, and leaves become dark green to purple in colour with a red leaf base.

contd

MINERAL DEFICIENCIES IN PLANTS contd

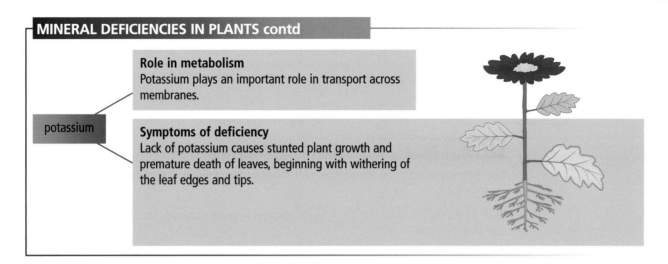

potassium

Role in metabolism
Potassium plays an important role in transport across membranes.

Symptoms of deficiency
Lack of potassium causes stunted plant growth and premature death of leaves, beginning with withering of the leaf edges and tips.

LET'S THINK ABOUT THIS

1. Why must oxygen be available to the roots if elements are to be absorbed?

2. In the table below, tick the boxes to indicate the correct appearance of plants that are deficient in certain macro-elements.

Macro-element	Symptom of deficiency			
	Leaf bases red	Overall growth reduced	Chlorotic leaves	Elongated root system
Nitrogen				
Magnesium				
Phosphorus				
Potassium				

LET'S THINK ABOUT THIS

Investigating the effect of mineral deficiency in cereal plants

The following method could be used to identify the effects of mineral deficiencies on plant growth.

1. A glass beaker is rinsed in concentrated nitric acid to remove any residue of mineral elements.

2. The outer surfaces of the beaker are painted black to prevent light from stimulating the growth of algae. Uncontrolled growth of algae would reduce the validity of the experiment.

3. A water culture solution is used, consisting of all essential minerals except the macro-element being studied.

4. Several oat seedlings are grown in each beaker to increase reliability.

5. Air is bubbled into the culture solution to provide roots with oxygen.

6. A control experiment is set up using a water culture solution with all essential minerals present.

7. After several weeks, the growth of the seedlings is examined and compared with the control group.

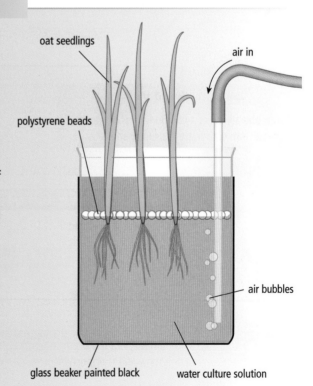

oat seedlings

air in

polystyrene beads

air bubbles

glass beaker painted black

water culture solution

ENVIRONMENTAL INFLUENCES ON ANIMALS

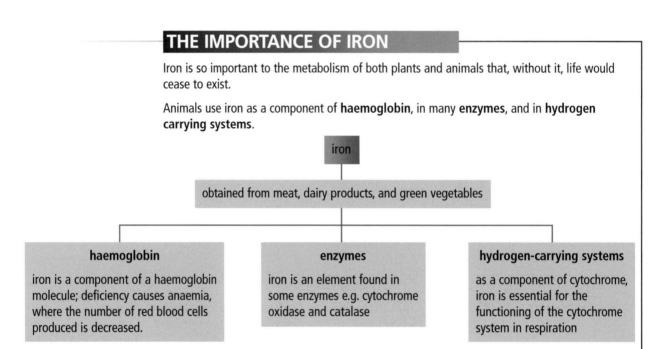

THE IMPORTANCE OF IRON

Iron is so important to the metabolism of both plants and animals that, without it, life would cease to exist.

Animals use iron as a component of **haemoglobin**, in many **enzymes**, and in **hydrogen carrying systems**.

iron

obtained from meat, dairy products, and green vegetables

haemoglobin

iron is a component of a haemoglobin molecule; deficiency causes anaemia, where the number of red blood cells produced is decreased.

enzymes

iron is an element found in some enzymes e.g. cytochrome oxidase and catalase

hydrogen-carrying systems

as a component of cytochrome, iron is essential for the functioning of the cytochrome system in respiration

THE IMPORTANCE OF CALCIUM

Calcium is found in **bones**, **teeth** and **shells**. It is also required for blood clotting.

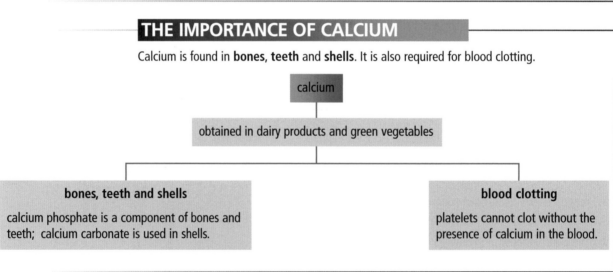

calcium

obtained in dairy products and green vegetables

bones, teeth and shells

calcium phosphate is a component of bones and teeth; calcium carbonate is used in shells.

blood clotting

platelets cannot clot without the presence of calcium in the blood.

IMPORTANCE OF VITAMIN D

Vitamin D is involved in **calcium** and **phosphate metabolism**.

vitamin D

obtained in oily fish, liver, eggs and through skin exposure to sunlight.

in humans, vitamin D promotes the absorption of calcium and phosphate during digestion and is important for bone growth and repair; deficiencies cause **rickets**.

EFFECTS OF DRUGS ON FOETAL DEVELOPMENT

Drugs taken during pregnancy can act on the placenta to reduce the exchange of nutrients and waste between foetal and maternal blood, or may cross the placenta to act directly on foetal cells.

thalidomide	Thalidomide is a drug that was taken by pregnant women to prevent morning sickness. If taken during the early stages of pregnancy, it crosses the placenta to act on cells of the limb buds, preventing the limbs from growing properly. Affected individuals have **limb deformities**.
alcohol	Alcohol can cross the placenta and cause the umbilical blood vessels to constrict, limiting the supply of oxygen to the foetus. If the mother has a high intake of alcohol, foetal alcohol syndrome can result, causing **mental disability**, **heart defects** and **poor growth**.
nicotine	Nicotine is a component of cigarette smoke. It causes the maternal blood vessels of the placenta to constrict, reducing exchange of oxygen and nutrients between mother and foetus. The foetal brain does not receive enough glucose and oxygen, causing **poor growth** and **mental development**.

 Look up www.chm.bris.ac.uk/motm/thalidomide/effects.html

⚙ LET'S THINK ABOUT THIS

Investigating the inhibiting effect of lead on enzyme function.

Lead is a metal that is poisonous to living organisms. This is due to its inhibitory effect on the enzymes involved in many metabolic pathways. We can use the reaction that causes browning in fruit to demonstrate this effect. When fruits such as apple and bananas are cut or bruised, the cells are damaged, allowing oxygen from the air to react with substances inside the cells. This causes the following reaction and colour changes to occur.

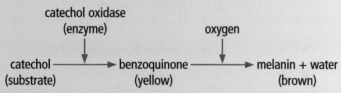

1. Apples are mashed and the juice is extracted by squeezing the tissue through a muslin bag. The juice contains the enzyme catechol oxidase.

2. Test tubes are set up containing catechol (substrate), buffer solution and either distilled water (control tube) or lead ethanoate (dilute or concentrated). Lead ethanoate is used, as the ethanoic acid produced in the reaction will not denature the enzyme.

3. Apple juice is added to each test tube and left to stand, allowing the reaction to occur.

4. Tube A (control, distilled water) turns dark brown as colourless catechol is converted to melanin. Tube B (dilute lead ethanoate) turns a light brown colour as some inhibition of enzyme action occurs. Tube C (concentrated lead ethanoate) does not change colour, as the reaction is completely inhibited by the lead ethanoate.

This experiment shows that:

- lead inhibits the action of the enzyme catechol oxidase
- the effect is increased when high concentrations of lead are present.

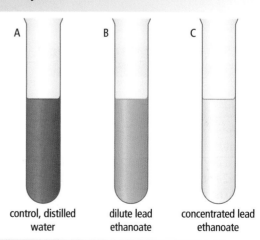

LIGHT

PHOTOPERIODISM

Photoperiodism is the biological response in organisms to changes in the **photoperiod** (the number of hours of daylight per day). Sexual reproduction in flowering plants and animals is affected by the photoperiod.

Flowering plants

Plants can be grouped according to the ratio of light to dark hours that they require for flowering to occur:

- short-day plants
- long-day plants
- day-neutral plants.

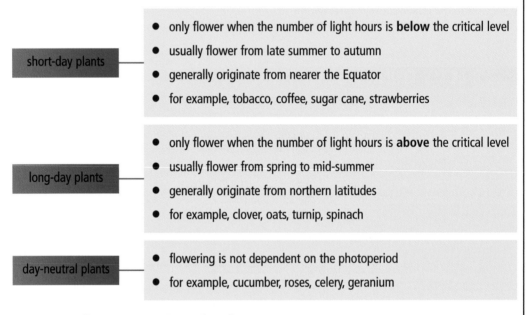

short-day plants
- only flower when the number of light hours is **below** the critical level
- usually flower from late summer to autumn
- generally originate from nearer the Equator
- for example, tobacco, coffee, sugar cane, strawberries

long-day plants
- only flower when the number of light hours is **above** the critical level
- usually flower from spring to mid-summer
- generally originate from northern latitudes
- for example, clover, oats, turnip, spinach

day-neutral plants
- flowering is not dependent on the photoperiod
- for example, cucumber, roses, celery, geranium

Breeding season in animals

Photoperiod determines the timing of the breeding season in many animals by controlling the production of sex hormones and gametes. In such animals, the testes and ovaries are only active at set times in the year.

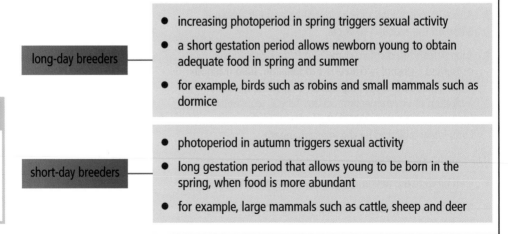

long-day breeders
- increasing photoperiod in spring triggers sexual activity
- a short gestation period allows newborn young to obtain adequate food in spring and summer
- for example, birds such as robins and small mammals such as dormice

short-day breeders
- photoperiod in autumn triggers sexual activity
- long gestation period that allows young to be born in the spring, when food is more abundant
- for example, large mammals such as cattle, sheep and deer

> **DON'T FORGET**
>
> Day length is **not** the same as photoperiod; a day length is always 24 hours, but photoperiod changes throughout the year.

LIGHT AND PHOTOTROPISM

The movement of a growing plant in response to light is called **phototropism** (photo = light, tropism = movement). Shoots grow towards the direction of a light source, due to the presence of **auxins** (see page 70). This allows the plant to maximise its chances of obtaining light for photosynthesis.

The table below compares the effects of high and low light levels on the growth of plants.

plant structure	Effect of light conditions	
	high light levels	**low light levels**

stem	strong	weak
internode	short	long
leaf	large, expanded, green	small, curled, yellow

When a plant is grown in darkness, it is said to be **etiolated.** By increasing stem length, the plant has more chance of obtaining the light that it needs for photosynthesis.

DON'T FORGET

Shoots show **positive phototropism** as they grow towards the light.

Look up http://www.britannica.com/EBchecked/topic/458258/phototropism

LET'S THINK ABOUT THIS

Migration and hibernation are other examples of photoperiodism that are of biological significance. The autumn migration of geese from Norway to Scotland is triggered by a decreasing photoperiod. Migrating to a food source allows the birds to survive through the winter.

HOMEOSTASIS – NEGATIVE FEEDBACK AND WATER BALANCE

Homeostasis is the **maintenance** of the body's **internal environment** in response to changes in the surroundings. Three key areas that require regulation are:

- water balance
- blood sugar level
- temperature.

NEGATIVE FEEDBACK

To allow homeostasis, there must be a corrective mechanism that acts when any variable in the internal environment changes too much. A mechanism like this uses **negative feedback**.

Aspects of the internal environment are monitored by **receptor cells** in **monitoring centres** around the body. Deviations from the normal level or **set point** (at which conditions are optimum for body processes) are detected by the receptor cells, and result in messages being sent from monitoring centres to **effector** organs. The effectors respond by bringing the level back to the set point. Messages sent out by the monitoring centres can be in the form of either:

- hormones that are secreted into the blood, or
- nerve impulses.

The diagram summarises the events of negative feedback.

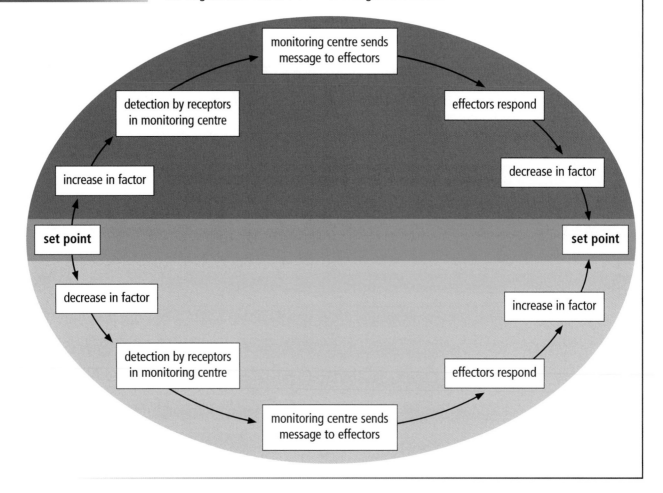

WATER BALANCE

The process of maintaining water and salt balance in the body is called **osmoregulation**. Osmoregulation uses negative feedback to maintain a constant water concentration in the body. The receptor cells (**osmoreceptors**) that detect the water concentration of the blood are found lining the blood vessels of the **hypothalamus**, the monitoring centre in the brain. From the osmoreceptors, nerve impulses pass to the **pituitary gland**, stimulating or inhibiting the production of **antidiuretic hormone (ADH)** from the pituitary gland.

In the case of low water concentration in the blood, ADH production is stimulated and the hormone travels in the blood to the kidney tubules. Here it acts on the ascending limb of the loop of Henlé, the distal convoluted tubule and the collecting ducts, making them more permeable to water. Water passes out of the tubule by osmosis and into the blood in the surrounding capillaries. If more **ADH** is produced, **less urine** is excreted.

The diagram below summarises the events of osmoregulation.

DON'T FORGET

Antidiuretic means **against** urine production, so **more ADH** gives **less urine**.

DON'T FORGET

You should revise the structure of a nephron from your Standard Grade notes.

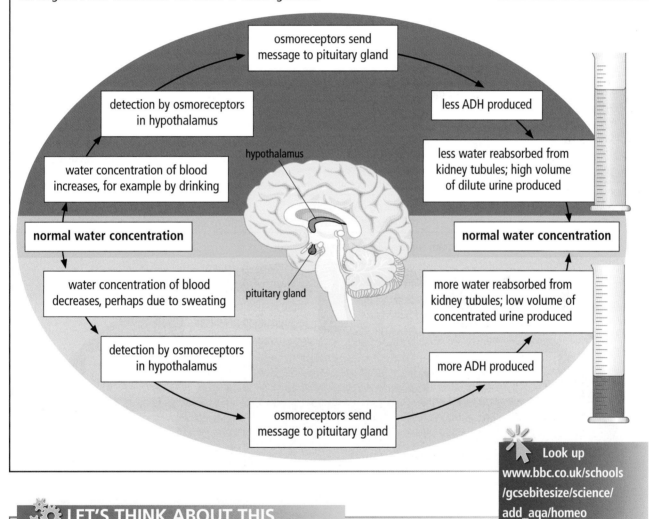

Look up
www.bbc.co.uk/schools
/gcsebitesize/science/
add_aqa/homeo

LET'S THINK ABOUT THIS

The water concentration of the blood is increased by drinking and by eating moist food. The water concentration of the blood decreases as a result of increased sweating, eating salty food, and through lack of drinking.

1. Where in the body is the monitoring centre for water concentration in the blood?

2. Which gland secretes ADH?

3. How does ADH reach the kidney tubules?

4. What is the effect of increased ADH secretion?

HOMEOSTASIS – BLOOD GLUCOSE LEVEL

Carbohydrates obtained in the diet are broken down into glucose. Glucose passes through the wall of the small intestine into the blood, which transports it to body cells. Glucose is used by body cells as the main respiratory substrate, and a constant supply is required to give energy in the form of ATP (refer back to page 18). Without homeostasis, the blood glucose level would be very high after a meal and become very low between meals; the required steady supply of energy could not be achieved.

Homeostasis promotes storage of glucose in the **liver** when there is excess in the blood, and stimulates release of glucose by the liver into the blood as body cells use it up in respiration. As a result of homeostasis, the blood glucose level is kept relatively constant. The **liver** is said to be a **storage reservoir** for **carbohydrates**.

DON'T FORGET

Do not get confused between the hormone **glucagon** and the storage carbohydrate **glycogen**. To remember the function of glucagon: when glucose is **gone**, you need gluca**gon**.

CONTROL OF BLOOD GLUCOSE LEVEL

Receptor cells in the **pancreas** monitor changes in blood glucose concentration. Depending on the glucose concentration detected, the pancreas produces one of two hormones.

Hormone	Produced in response to	Effect
insulin	increased blood glucose concentrations after eating a meal	liver cells respond by converting glucose into glycogen – blood glucose levels decrease
glucagon	decreased blood glucose concentrations between meals	liver cells respond by converting glycogen into glucose – blood glucose levels increase

The diagram summarises the steps involved.

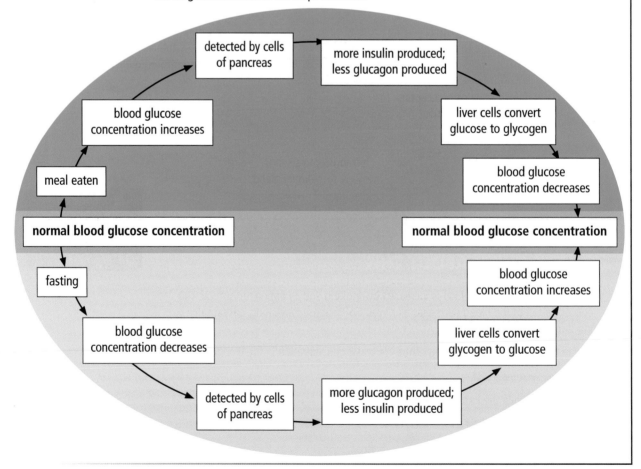

ADRENALINE

The hormone **adrenaline** is produced by the adrenal glands in times of stress and is involved in preparing the body for the 'flight or fight response'. In the presence of adrenaline, liver cells convert glycogen into glucose, increasing the blood glucose concentration above the normal level. This provides muscles with the glucose that they require during increased levels of activity.

LET'S THINK ABOUT THIS

The diagram below shows some of the stages in the control of a person's blood glucose level.

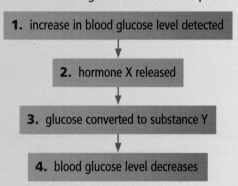

1. increase in blood glucose level detected

2. hormone X released

3. glucose converted to substance Y

4. blood glucose level decreases

(a) In which organ would the increase in blood glucose level be detected?

(b) Name hormone X and substance Y.

(c) Where does stage 3 occur?

(d) If hormone X was not produced, the blood glucose level would drop very slowly. Why?

LET'S THINK ABOUT THIS

The blood glucose concentration constantly fluctuates around the normal level (set point), as shown in the graph below.

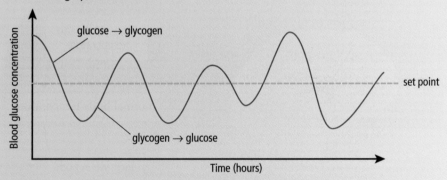

When the blood glucose level drops below the set point, less insulin and more glucagon is produced, causing the blood glucose level to rise again.

When the blood glucose level goes above the set point, more insulin and less glucagon is produced, causing the blood glucose level to decrease again.

HOMEOSTASIS – TEMPERATURE

All chemical reactions in the body are controlled by enzymes. When body temperature is below the optimum temperature for enzyme function, metabolism is slow. Metabolism is fastest when the body temperature is close to the optimum temperature for enzymes. Above the optimum temperature, enzymes start to denature; the metabolism slows down and eventually stops.

Animals can be divided into two groups according to their ability to regulate body temperature:

- **Ectotherms** have no homeostatic control and derive body heat from their surroundings. Examples are fish, amphibians and reptiles.

- **Endotherms** have homeostatic control. A high metabolic rate allows endotherms to produce heat via their own metabolism, meaning that their body temperature is independent of the temperature of the surroundings. Examples are mammals and birds.

REGULATION OF BODY TEMPERATURE IN ENDOTHERMS

Receptors that detect blood temperature (**thermoreceptors**) are found in the lining of blood vessels in the **hypothalamus** – the **temperature-monitoring centre** in the brain. When thermoreceptors detect changes in blood temperature, the hypothalamus sends out nerve impulses to effector organs in the skin and body muscles. The effectors bring about a response designed to return the temperature to normal.

The diagram summarises the process.

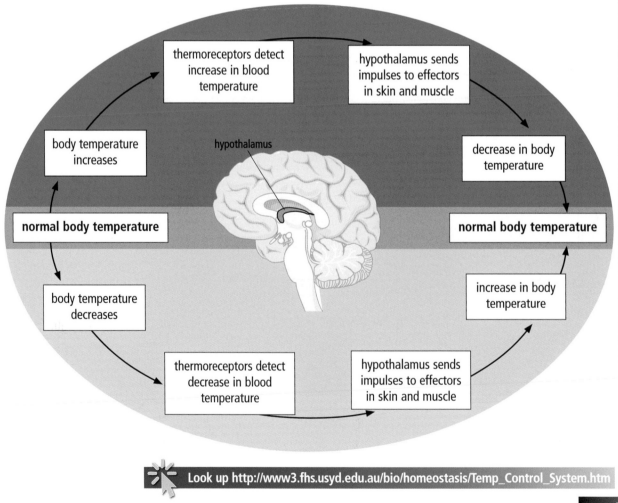

Look up http://www3.fhs.usyd.edu.au/bio/homeostasis/Temp_Control_System.htm

contd

REGULATION OF BODY TEMPERATURE IN ENDOTHERMS contd

Involuntary response to temperature change in endotherms

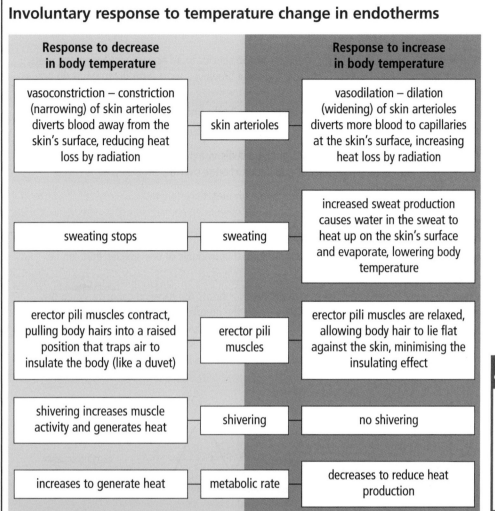

Response to decrease in body temperature		Response to increase in body temperature
vasoconstriction – constriction (narrowing) of skin arterioles diverts blood away from the skin's surface, reducing heat loss by radiation	skin arterioles	vasodilation – dilation (widening) of skin arterioles diverts more blood to capillaries at the skin's surface, increasing heat loss by radiation
sweating stops	sweating	increased sweat production causes water in the sweat to heat up on the skin's surface and evaporate, lowering body temperature
erector pili muscles contract, pulling body hairs into a raised position that traps air to insulate the body (like a duvet)	erector pili muscles	erector pili muscles are relaxed, allowing body hair to lie flat against the skin, minimising the insulating effect
shivering increases muscle activity and generates heat	shivering	no shivering
increases to generate heat	metabolic rate	decreases to reduce heat production

DON'T FORGET

Body temperature can increase due to exercise, illness and exposure to a hot environment. Body temperature can decrease due to exposure to a cold external environment.

LET'S THINK ABOUT THIS

1. Where in the body is the temperature-monitoring centre?
2. How does the temperature-monitoring centre communicate with the effectors?
3. Why is body temperature important in carrying out metabolic processes?

LET'S THINK ABOUT THIS

Changes in surface temperature on the body can be measured using a thermistor. You may have carried out an investigation in which thermistors were placed in the armpit (representing the body core) and between the fingers of one hand (representing the body shell), while the other hand was held in a bucket of icy water. Receptors in the skin detect a drop in temperature in the hand in the icy water and send nerve impulses to the temperature monitoring centre in the hypothalamus. The hypothalamus responds by stimulating vasoconstriction in both hands, diverting blood away from the body surface to conserve heat. A drop in temperature would be registered in the thermistor held between the fingers, as heat is conserved. The thermistor held in the armpit would register a constant temperature.

POPULATION DYNAMICS

A **population** consists of all the organisms of one species in an ecosystem. Over many years, the size of a population remains broadly the same. However, short-term fluctuations do occur, depending on:

- the number of newcomers to the population – births and immigrants
- the number leaving the population – deaths and emigrants.

The **population density** is the number of individuals in a population within each unit area or volume of the habitat. The study of population change over a period of time is called **population dynamics**.

CARRYING CAPACITY

The **carrying capacity** is the maximum number of individuals of one species that can be supported within a habitat.

Often, a population will increase in size until it overshoots the carrying capacity. When this happens, the environment can no longer support the population; lack of food, overcrowding, predation and increased incidence of disease cause the population size to decrease. When the population falls below the carrying capacity, the environment is able to recover. This, in turn, allows more individuals to survive, and the population size increases until it overshoots the carrying capacity, and the cycle begins again.

DON'T FORGET

A **dynamic equilibrium** is established when birth rate equals death rate. Population size is stable as a result.

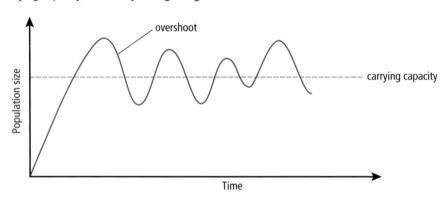

 Look up http://ats.doit.wisc.edu/biology/ec/pd/t2_a1.htm

FACTORS AFFECTING POPULATION SIZE

The factors that affect population size can be divided into two groups:

- density-independent factors
- density-dependent factors.

Density-independent factors

These factors affect populations no matter what the size, and the percentage of the population that is affected is the same for a small population as it is for a large population.

Density-independent factors include:

- extremes of rainfall (drought and flood)
- spells of extremes of temperature
- forest fires
- storms.

contd

FACTORS AFFECTING POPULATION SIZE contd

Density-dependent factors

An increase in population density puts pressure on resources within the ecosystem. As the density increases, there is a reduction in the resources available for each individual. Eventually, resources become so limited that any further increase in population size is prevented.

Examples of density-dependent factors are shown in the diagram.

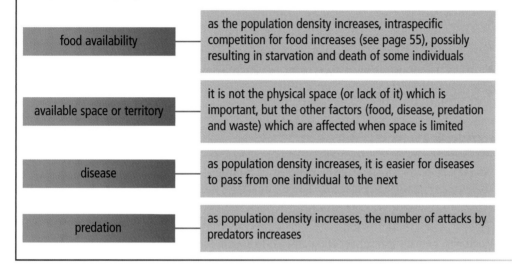

food availability	as the population density increases, intraspecific competition for food increases (see page 55), possibly resulting in starvation and death of some individuals
available space or territory	it is not the physical space (or lack of it) which is important, but the other factors (food, disease, predation and waste) which are affected when space is limited
disease	as population density increases, it is easier for diseases to pass from one individual to the next
predation	as population density increases, the number of attacks by predators increases

LET'S THINK ABOUT THIS

Predator–prey interactions

You should be able to explain the shape of a predator–prey density graph. The predator density mimics the prey density, but lags behind it. This is because it takes time for changes in the number of prey to have an effect on the number of predators.

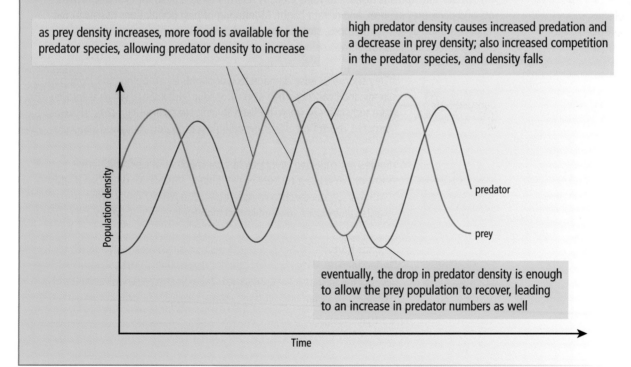

as prey density increases, more food is available for the predator species, allowing predator density to increase

high predator density causes increased predation and a decrease in prey density; also increased competition in the predator species, and density falls

eventually, the drop in predator density is enough to allow the prey population to recover, leading to an increase in predator numbers as well

predator

prey

Population density

Time

MONITORING POPULATIONS

It is very important that we monitor populations if we are to protect and enhance our resources and preserve the world's ecosystems.

Data that is gathered allows us to:

- manage species used for food or raw materials
- control pest species
- detect changes in pollution levels using indicator species
- protect endangered species.

FOOD AND RAW MATERIALS

If the rate at which individuals are removed from a population is too high, they cannot be replaced fast enough. Population size may fall below a critical level, eventually resulting in extinction of the species. By monitoring population size, the **maximum sustainable yield** (the rate at which individuals can be harvested without affecting future productivity of the species) can be determined.

Monitoring of many fish species allows the EU to impose fishing quotas which ensure the sustainable management of fish stocks. Fishing quotas represent a compromise between fishermen, who need to make a living, and environmentalists, whose data would suggest that some fish populations are close to collapse. In December 2008, quotas for cod and plaice in the North Sea were increased, whereas a total ban on fishing for anchovies in the Bay of Biscay has been established to allow the population to recover.

CONTROL OF PEST SPECIES

Any species that spreads disease or ruins crops is referred to as a pest, for example locusts, rats and mice, grey squirrel, and potato blight. Monitoring of pest populations assists in the implementation of control methods, such as the use of pesticides or biological methods of control. Three examples are shown below.

grey squirrel	Grey squirrels cause damage to woodlands by stripping bark from tree stems and branches. They also carry squirrelpox virus, which is fatal to rare red squirrels. Traps are used to indicate squirrel numbers, allowing targeted control measures including warfarin poisoning and shooting.
grey mould in plants	Spores are collected on plates of agar medium placed at different locations in a plant nursery or polytunnel. This allows detection of the pathogen before symptoms develop on the plants. Visual inspection of plants can be used to identify and monitor infection if symptoms develop.
aphids	Aphids transmit plant viruses and also cause physical damage to plants when feeding. Winged aphids can be monitored using sticky traps to collect the insects for identification. The presence of wingless aphids is assessed by visual inspection of crops.

DETECTION OF POLLUTION LEVELS USING INDICATOR SPECIES

An **indicator species** is a species whose **presence** or **absence** gives an indication of the level of pollution.

lichen

lichen is an indicator species for **sulphur dioxide pollution** in the air; where high levels of sulphur dioxide are present, lichen will be absent from the ecosystem

Some invertebrates are indicator species for levels of **dissolved oxygen** in fresh water.

rat-tailed maggot

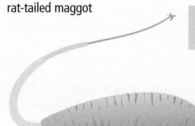

rat-tailed maggots and **sludge worms** are found in polluted water with low levels of dissolved oxygen

mayfly nymph

mayfly nymphs are found where dissolved oxygen levels are high and pollution is low

Look up
www.epa.gov/bioiweb1/
html/indicator.html

Look up
www.earthdive.com/site/
ecoregions/default.asp

PROTECTION OF ENDANGERED SPECIES

An **endangered species** is a species that is in danger of extinction either because its numbers are below a sustainable level or because its habitat is under threat. For example, the **polar bear** is classed as a vulnerable species due to the decline in population size over several generations; and the **Iberian Lynx** is a critically-endangered species with a population size of approximately 150.

Endangered species can be protected in a number of ways, including:

- establishment of wildlife and nature reserves
- conservation of habitats
- captive breeding programmes
- cell and seed banks
- international agreements, hunting bans, quotas, other legislation and laws.

Monitoring populations of endangered species allows methods of conservation to be evaluated and reviewed.

Look up
www.worldwildlife.org/
species/

Look up
www.fws.gov/endangered/

LET'S THINK ABOUT THIS

Protection of whales

There are over 80 species of cetaceans (whales, dolphins and porpoises). 13 of these are species of great whale. Seven of the great whale species are either endangered or vulnerable. Why?

Pollution, climate change, habitat destruction, and of course commercial whaling, have all contributed to the decline in whale numbers. Commercial whaling alone results in the death of over 1000 whales every year - although commercial whaling is actually banned under an international moratorium.

SUCCESSION AND CLIMAX IN PLANT COMMUNITIES

SUCCESSION

Succession is the unidirectional and gradual change in the diversity of plant species in a habitat. It is accompanied by habitat modification.

- **Primary succession** occurs in areas that were previously uninhabited.
- **Secondary succession** is the recolonisation of areas left barren either by natural means (such as forest fires) or by artificial means (for example, deforestation).

PRIMARY SUCCESSION

The first species to colonise a new area are called **pioneer species**. Pioneer species are usually wind-dispersed plants and are adapted to survive adverse conditions, such as **lichens and mosses**. They have shallow root systems and a low nutrient requirement.

As pioneer plants die and fall to the ground, they form a layer of humus (dead organic matter). This decomposes through the action of bacteria and fungi, and improves the soil by:

- providing air spaces
- increasing depth
- improving water retention and drainage
- improving pH
- increasing fertility (increasing the availability of nutrients).

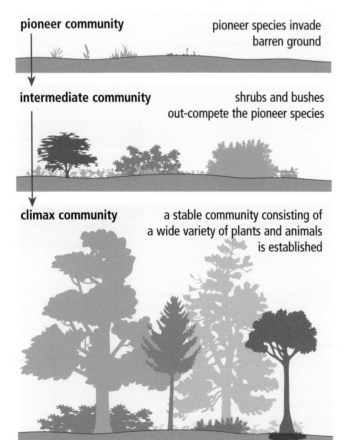

pioneer community   pioneer species invade barren ground

intermediate community   shrubs and bushes out-compete the pioneer species

climax community   a stable community consisting of a wide variety of plants and animals is established

Over time, the habitat is modified so much that it becomes favourable to other species. **Intermediate species** such as **large grasses and shrubs** enter and colonise the habitat, out-competing and succeeding the pioneer plants.

As more plants grow and die, soil quality continues to improve.

Succession continues with larger and larger plant species colonising the area, until a **climax community** has evolved. Climax communities are very stable and can survive without change, as long as there is no alteration to the climate. It can take thousands of years to progress from barren habitat to climax community.

SECONDARY SUCCESSION

Secondary succession takes place in a cleared habitat that was previously colonised. If a nutrient-rich soil that can support intermediate communities remains, colonisation takes less time to reach a climax than in primary succession (perhaps 100 years, compared to thousands of years).

CLIMAX COMMUNITIES

Climax or final communities show certain characteristics.

more complex food webs	as the number of different plant species is diverse, a greater variety of animals can be supported, particularly insects, leading to more interconnections between species in the food web
greater biomass	plant biomass increases as succession progresses, peaking in climax communities where the soil is deep and rich enough in organic matter and minerals to support a large biomass of plants
greater species diversity	better-quality soils support a greater variety of plant species, in turn providing food and habitats for many different animals
greater stability	a climax community matures over time but is not replaced by further communities; it is said to be **self-perpetuating**

LET'S THINK ABOUT THIS

Sand dunes are good examples of succession. The sequence of events that leads to formation of a climax community on a sand dune is as follows:

1. Pioneer plants, such as sea holly, colonise the area just above the high tide level.

2. Sand becomes trapped between the pioneer plants, forming sand dunes.

3. Couch grass grows on the new sand dune, sending out rhizomes that stabilise the sand.

4. When the dune becomes high enough to prevent immersion in sea water, marram grass is able to move in. Marram grass acts as a habitat for small animals.

5. Animal and plant remains build up, forming humus. This improves the soil quality, allowing other plant species to colonise the dune.

increasing age of sand dune

←

climax community   marram grass   couch grass   pioneer plants

sea

EXTENDED-RESPONSE QUESTIONS

ESSAY GUIDE

When writing essays for the Higher Biology exam, there are some easy-to-follow guidelines that will help you to maximise your grade.

- Within each essay question, you have the choice of two titles. Don't jump at the essay that you think looks easier. Make a very brief plan for each title to make sure you pick the one that will give you more marks.

- Use your essay plan to ensure that you stick to the question being asked in the title. You can waste an awful lot of time writing down irrelevant information that will not gain marks.

- Biology essays should be an organised list of facts, placed in the correct context. So, don't waste time writing an opening paragraph (as you would in English); get straight to the point.

- In Essay 1, the marks allocated to each section of the essay are given on the paper. These marks are purely for knowledge. It is a good idea to write the facts in an ordered bullet-point list under the subheadings given in the essay title.

- Essay 2 has eight marks allocated for knowledge. An additional one mark is awarded for relevance and one mark for coherence (flow). If you are short of time, it may be wise to bullet-point this essay and not worry about the coherence mark.

- Spelling and grammar are not being marked. As a general rule, if you spell the word as it would be said, so that it cannot be mistaken for any other biological term, the mark will be awarded.

- Diagrams are a very useful way of displaying a lot of information, but they can take time to draw. If you are using diagrams, they should be quick sketches and must be labelled. If you are adding arrows to diagrams, they should point in the right direction and must have arrowheads. Do not colour in diagrams, as this wastes time and will give no extra marks.

- Finally, do not score out your essay plans. The examiner will look for marks in everything you have written, and any facts that you had placed in your plan but missed in your essay will be credited.

EXAMPLES OF EXTENDED-RESPONSE ESSAY ANSWERS

Essay 1

Give an account of homeostasis under the following headings:

(a) Control of water balance   **(5 marks)**

(b) Hormonal control of blood glucose level   **(5 marks)**

Control of water balance

- *Osmoreceptor cells in the hypothalamus detect changes in water concentration in the blood.*
- *Nerve impulses send information from the hypothalamus to the pituitary gland.*
- *If the water concentration in the blood is too low, the pituitary gland produces more antidiuretic hormone (ADH).*
- *If the water concentration in the blood is too high, the pituitary gland produces less ADH.*

contd

EXAMPLES OF EXTENDED-RESPONSE ESSAY ANSWERS contd

- ADH travels in the blood to the nephrons in the kidney.
- Increased levels of ADH cause the ascending limb of the loop of Henlé, distal convoluted tubule and collecting duct to be more permeable to water.
- This increases water reabsorption, and a low volume of concentrated urine is produced.

Hormonal control of blood glucose concentration

- After a meal, the blood glucose level increases.
- Receptor cells in the pancreas detect blood glucose concentrations.
- High blood glucose levels cause the pancreas to secrete the hormone insulin.
- Insulin travels in the blood to the liver and causes liver cells to convert glucose into glycogen.
- This brings the blood glucose level back down to normal again.
- Between meals, the blood glucose level decreases.
- Receptor cells in the pancreas detect this decrease and produce the hormone glucagon.
- Glucagon passes in the blood to the liver and causes liver cells to convert glycogen to glucose.
- This increases the blood glucose level again.

Essay 2

Give an account of the homeostatic control of water balance and glucose concentration.

Negative feedback maintains the optimum levels of water and glucose in the body. Blood water concentration is detected by osmoreceptors in the hypothalamus. The hypothalamus sends nerve impulses to the pituitary gland. When there is a low water concentration in the blood, the pituitary gland is stimulated to produce more antidiuretic hormone (ADH). ADH passes in the blood to the kidneys, where it acts on cells of the ascending limb of the loop of Henlé, distal convoluted tubule and collecting duct, making the cells more permeable to water. This causes more water to be reabsorbed, and a low volume of concentrated urine is produced. The water concentration of the blood is returned to normal. When the water concentration of blood goes too high, less ADH is produced, making the kidney tubule less permeable to water. A high volume of dilute urine would then be produced.

Glucose concentration is detected by cells of the pancreas. After a meal, the blood glucose level increases above normal, and the pancreas is stimulated to produce the hormone insulin. Insulin travels in the blood to the liver, where it causes liver cells to convert glucose into glycogen. The blood glucose level now returns to normal. When the blood glucose level decreases below normal, as it does between meals, the pancreas is stimulated to produce the hormone glucagon. Glucagon travels in the blood to the liver. Liver cells now convert glycogen into glucose, which is released into the blood, raising the blood glucose level back to normal.

PROBLEM-SOLVING

DEVELOPING THE RIGHT SKILLS

Problem- solving involves assessing and processing information that has been given to you to interpret. By mastering these skills, you will be maximising your chances of a good pass in the exam. You will benefit from using past papers to practise problem-solving questions, enhancing your technique and identifying areas that require further revision. The advice below covers the key problem-solving skills as defined in the Higher Biology arrangements.

SELECTING RELEVANT INFORMATION

- Often, you will be asked to select information from a graph. Read the x- and y-axes carefully, taking note of the variables and units being used. Use a ruler to help you read values from both axes – be accurate.
- If two sets of values have been plotted with different y-axes, check for differences in scale. Be careful to read the correct y-axis for each line or set of values.
- Make sure that you read the question properly, underlining any relevant information.

PRESENTING INFORMATION

- It is likely that you will be asked to take information from a table and present it in the form of a graph.
- Use the column headings from the table as the axis labels on your graph. Copy the headings exactly as they have been written, including units.
- The variable that has been deliberately altered in the experiment should be placed on the x-axis. This is usually the set of data in the left column of the table. The variable that has been obtained as the result goes on the y-axis of the graph.
- Your graph must cover 50% or more of the graph paper. Decide on a suitable scale, making sure that you can plot all the points. The scale should be linear – each division must be worth the same number of units; that is, the number of squares between 10 and 20 must be the same as the number of squares between 20 and 30 on the same axis.
- **Bar graphs**: bars should be of equal width, using a ruler to draw a line around the boundary of the bar. You may be asked to include two sets of data and so should distinguish between sets by shading bars and including a key.
- **Line graphs**: mark points with a lightly-drawn X and join points together with a fine line drawn using a ruler. Do not extend the line to the zero point on the x-axis **unless** this value has been included in the table, and make sure that the line does not extend beyond the last point given in the table.
- Don't panic if you make a mistake that you cannot correct easily. The exam paper includes an extra grid at the end of the paper.

PROCESSING INFORMATION

Averages

Add up all values and divide by the total number of values.

For example, the average of 4, 6 and 8 is $\frac{4 + 6 + 8}{3} = 6$

Ratios

To express ratios correctly, write down the raw values in the order that the question suggests. Then try to find a number that divides into both values. The final answer must be in as simple a form as possible, so you may have to try dividing the values by several numbers.

PROCESSING INFORMATION contd

For example, if the number of seeds produced by plant A was 25 and the number of seeds produced by plant B was 35, the ratio of seeds produced by plant B to seeds produced by plant A would be as follows.

35:25 → divide by 5 → 7:5

There is no number that both 7 and 5 can be divided by, so the answer is 7:5.

Percentages

To calculate a percentage, the value that is to be expressed as a percentage is divided by the total of all values, and the resulting number is multiplied by 100. For example, if 300 eggs were fertilised but only 50 of them hatched, the percentage that survived would be:

$$\frac{value}{total} \times 100 = \frac{50}{300} \times 100 = 16 \cdot 7\%$$

To calculate the exact number when the value is expressed as a percentage of the whole, take the percentage, divide by 100 and multiply by the total number.

For example, if 60% of seeds germinated out of a total of 200, the number of seeds that germinated was:

$$\frac{percentage}{100} \times total = \frac{60}{100} \times 200 = 120$$

Percentage change

To calculate percentage change (either increase or decrease), work out the difference between the initial and final value, divide by the initial value and then multiply by 100.

For example, if a pulse rate before exercise is 70, rising to 120 after exercise, the percentage increase would be:

$$\frac{difference}{initial\ value} \times 100 = \frac{120 - 70}{120} \times 100 = 41 \cdot 7\%$$

PLANNING, DESIGNING AND EVALUATING EXPERIMENTS

- Experiments are repeated to give a range of results. An average is calculated to increase **reliability**.
- Only one variable should be altered in each experiment. All other variables must be kept constant to ensure that the experiment is **valid**. (Do use the word 'valid' in your answer; do not use the word 'fair'.)
- Control experiments should be used to ensure that any change in results can be attributed to the variable that was altered. Suitable controls may be to replace organisms with the same mass or volume of glass beads, to use boiled (inactive) tissue or enzymes, or to replace solutions with the same volume of distilled water.
- When identifying variables that should be kept constant, read the question carefully. Usually, you are required to give variables 'not already mentioned' in the question. Good examples to think about are temperature, mass or volume of substances. (Do use the correct term for the variable in your answer; do not use the word 'amount'.)
- Where initial values were not identical, the percentage change must be used to compare results between groups.

DON'T FORGET

Repeats give **reliability**; variables control **validity**.

DRAWING CONCLUSIONS AND MAKING PREDICTIONS

- When writing a conclusion, describe the overall trend in results. You must include discussion of both variables in your answer. Make sure you describe completely any observed trend. If a graph shows results increasing and then levelling off or decreasing, you must describe both parts of the graph. The clue is in the number of marks assigned to the question: if it is worth two marks, there must be at least two parts to the graph to describe!
- If asked to predict results from a graph, use a ruler to draw a line that extends the graph on the graph paper.

ANSWERS

CELL BIOLOGY

p.7

1 (a)

Part of cell	Name	Function
B	mitochondrion	site of aerobic respiration
E	Golgi apparatus	packages proteins for secretion
C	RER	transports protein

(b) Microvilli increase surface area for absorption of substances.

p.15

1. carbon dioxide: 1   RuBP: 5   GP: 3   glucose: 6
2. ATP and hydrogen needed to convert GP $\rightarrow$ triose phosphate
3. photolysis: grana carbon fixation: stroma of chloroplast

p.17

1. Pre-incubation in a warm water bath ensures that temperature variable is kept constant, increasing validity.
2. Yeast suspension is prepared two days before the experiment to give time for yeast to use up any stored carbohydrate.

p.21

1.

Statement	Stage of aerobic respiration			Anaerobic respiration
	Glycolysis	Kreb's cycle	Oxidative phosphorylation	
oxygen required		✓	✓	
carbon dioxide produced		✓		
hydrogen ions attach to NAD	✓	✓		✓
occurs in the matrix of the mitochondria		✓		
occurs in the cytoplasm	✓			✓
occurs on the cristae of the mitochondria			✓	
produces ATP	✓		✓	✓
ethanol or lactic acid produced				✓

2. Oxygen debt is the amount of oxygen needed to complete the breakdown of lactic acid produced in anaerobic respiration.
3. oxygen
4. glucose: 6   pyruvic acid: 3   acetyl co-enzyme A: 2   citric acid: 6

p.23

(a) fibrous
(b) collagen - long parallel chains provide a strong, rope-like structure

(c) conjugated proteins have non-protein substances bound to the protein
(d) The shape of the enzyme includes an active site which is complementary in shape to the substrate on which it acts.

p.29

1 (a) mRNA codons: AUG   CAU   CGG   AGU
    tRNA anticodons: UAC   GUA   GCC   UCA
    amino acid sequence:
    methionine   histidine   arginine   serine
    (b) (i) AAG   (ii) AUU
    (c) UCC and GCC and GCA

GENETICS AND ADAPTATION

p.39

1. Cross 1:

	X^n	Y
X^N	$X^N X^n$	$X^N Y$
X^n	$X^n X^n$	$X^n Y$

F_1 genotypes: $X^N X^n$ $X^n X^n$ $X^N Y$ $X^n Y$

F_1 phenotypes: carrier female   colour-blind female   normal male   colour-blind male

Cross 2:

	X^N	Y
X^N	$X^N X^N$	$X^N Y$
X^n	$X^N X^n$	$X^n Y$

F_1 genotypes: $X^N X^N$ $X^N X^n$ $X^N Y$ $X^n Y$

F_1 phenotypes: normal female   carrier female   normal male   colour-blind male

2. $X^H X^H$ normal female $X^H X^h$ carrier female
   $X^h X^h$ haemophilic female $X^H Y$ normal male
   $X^h Y$ haemophilic male

p.41

1. (a) substitution   (b) deletion   (c) inversion
2. One or more bases have been removed from the sequence. This can happen naturally during DNA replication or as a result of a mutagenic agent such as radiation.

p.43

1. An homologous pair fails to separate during meiosis, resulting in one daughter cell receiving one less chromosome and the other daughter cell receiving one too many chromosomes.
2. Polyploidy makes plants more fertile, grow more successfully and have more disease resistance.

p.49

(a) B, D, E, C, A
(b) endonuclease
(c) by using gene probes and banding pattern recognition
(d) Unwanted characteristics are kept to a minimum, takes less time to produce the desired characteristics and new combinations of characteristics can be achieved.

p.51

1.

Organism	Why adaptation is required	Kidney adaptation	Urine production
desert mammal	Animal loses water to the environment	long loop of Henle	low volume
fresh-water fish	gain water constantly from environment	many, large glomeruli	large volume

2. A, C, F

p.61

1. reduces damage from grazing animals
2. stinging nettles hairs, when broken, inject irritant into the skin
3. presence of thorns that cause damage to mouth of a grazing animal
4. underground storage organs that can grow roots and shoots, allowing rapid spread of plant
5. low meristems allow the plant to regrow very rapidly from buds at the base

CONTROL AND REGULATION

p.65

1. (a) false   (b) true   (c) false   (d) true
    (e) true   (f) true
2. Repressor molecule binds with lactose (inducer) rather than operator. This allows the operator to 'switch on' the structural gene, leading to transcription and translation of the genetic code for -galactosidase.

p.67

1. A mistake in the base sequence of a gene, preventing an enzyme being made and blocking a metabolic pathway.
2. (i) Phenylalanine is broken down by the mother (who does make the enzyme).
    (ii) Baby obtains phenylalanine through diet but cannot break it down, so it accumulates.
3. By switching genes off there is no waste of either energy or resources.

p.69

1. (i) thyroid gland   (ii) pituitary gland   (iii) pituitary gland
2. bones and soft tissues
3. thyroxine increases metabolic rate
4. GA is released into the aleurone layer, where it stimulates -amylase production.
5. -amylase catalyses breakdown of starch into maltose which can be used in respiration.

p.73

1. Oxygen is required for respiration to provide energy for absorption
2.

Macro-element	Symptom of deficiency			
	Leaf bases red	Overall growth reduced	Chlorotic leaves	Elongated root system
nitrogen	✓			✓
magnesium		✓	✓	
phosphorus	✓	✓		
potassium		✓		

p.79

1. hypothalamus
2. pituitary gland
3. ADH passes into the bloodstream
4. Increased ADH secretion causes increased reabsorption of water from the kidney tubule.

p.81

(a) pancreas
(b) X – insulin Y – glycogen
(c) in the liver
(d) glucose is used in respiration

p.83

1. hypothalamus
2. nerve impulses
3. If body temperature falls too low, metabolism will be too slow to maintain life. If body temperature goes too high, enzymes are denatured and metabolism stops.

INDEX

absorption 8–9
absorption spectra, for leaf
  pigments 11
actin 22
action spectra, for leaf
  pigments 11
active transport 9
adaptation 50, 53
adaptive radiation 45
adenine 24, 25, 26
adenosine diphosphate (ADP) 16
adenosine triphosphate (ATP) 12,
  16, 17
adhesion forces 52
ADH 79
ADP 16
adrenaline 81
aerobic respiration 20, 21
alcohol 75
aleurone layer 69
alleles 35, 36, 38
α-amylase 69
amino acids 22, 27
amylase 23
anaerobic respiration 18, 20, 21
anaphase I 34
anaphase II 35
antibodies 23, 31
anticodon 27
antidiuretic hormone (ADH) 79
antigens 31
ants 54
aphids 86
apical dominance 70
apical meristems 62, 70
artificial selection 46–9
ATP 12, 16, 17
autosomes 42
autotrophicity 54, 56
auxins 71
averages 92
avoidance behaviour 58

base sequences 24, 27
bases, pairing 27
bats 54
bees 54
behaviour 54, 55, 58, 59
β-galactosidase 64, 65
birds 44, 59
blood clotting 74
blood glucose levels 80, 81
body temperature 82–3
breeding season 76

calcium 74
Calvin cycle 13
cambium cells 63
camels 51
carbon fixation 13, 14, 15
carotene 10, 11
carriers 39
carrying capacity 84
catalase 74
catechol oxidase 75
cell differentiation 67
cell membrane proteins 8, 23
cell membranes 6
  active transport across 9
cell nucleus 6

cell walls 6
cells, ultrastructure 6
cellular defence mechanisms
  30–1, 32–3
cellular organisms 6–7
centromeres 34
channel-forming proteins 8
chiasma(ta) 34, 35
chimpanzees 59
chlorophyll 10, 11, 72
chloroplasts 6, 12
chlorosis 72
chromatids 34
chromatography 10
chromosomes 35
  homologous 34, 35, 42
  mutations 41, 42–3
  X 34, 38
  Y 38
cillia 6
climax communities 88, 89
Clostridium difficile 45
codons 26, 27
cohesion forces 52
collagen 22, 28
colour blindness 38–9
compensation points 57
competition 55, 56
competition exclusion
  principle 55
conclusions 93
conjugated proteins 23
co-operative hunting 55
cristae 16
crossing over 34, 35
cyanide 32
cyanogenic plants 32
cytochrome oxidase 74
cytochrome system 16, 20, 74
cytokinesis 35
cytoplasm 6, 18, 26, 27
cytosine 24, 25, 26

Darwin's finches 44
day-neutral plants 76
defence mechanisms
  animals 59
  cells 30–1, 32–3
  plants 60
dehydrogenase 17
deletion 40, 41
deoxyribonucleic acid (DNA) 24,
  25, 26
diffusion 8
dihybrid cross 36, 37
diploid 34
DNA 24, 25, 26
dolphins 54
dominance hierarchy 55
dominant genes 37, 38
Down's Syndrome 42, 43
Drosophila 36
duplication 41

E. coli 64
ecological isolation 44
ectotherms 82
effector organs 78
elephants 59
endangered species 87

endocrine glands 68
endonucleases 49
endotherms 82–3
energy release, in respiration 16
enzymes 23, 28, 66, 74
  β-galactosidase 64, 65
  catalase 74
  catechol oxidase 75
  cytochrome oxidase 74
  endonucleases 49
  ligase 49
erector pili muscles 83
Escherichia coli (E. coli) 64
essay guidelines 90
etiolation 77
Euglena 6
experiments, planning, designing
  and evaluating 93
extended-response essay answers,
  examples 90–1

fats 20
fibrous proteins 22
final communities see climax
  communities
fish 50–1
fishing quotas 86
flagellum 6
flowering 76
fluid mosaic model 8
foetal development, effect of
  drugs 75
food, obtaining 54–7
foraging behaviour 54
frame-shift mutations 40
fruit bruising 75

gametes 34, 35, 36, 42
gene pools 44
genes
  linked 36–7
  mutations 40, 41
  plant height and flower
    colour 36
  sex-linked 38–9
  switched on/off 64, 67
genetic engineering 48–9
genetic variation 34
geographical isolation 44
gibberellic acid (GA) 69
globular proteins 22, 23
glucagon 80
glucose 18, 21, 80
glycolysis 18
goblet cells 7
Golgi apparatus 6, 28
grana 12
graphs 92, 93
grazing levels 56
grazing toleration 61
growth 62–3
growth hormone (GH) 68
guanine 24, 25, 26
guard cells 7, 53

habituation 58
haemoglobin 23, 74
haploid 34
heterotrophicity 54, 56
hierarchy 55

holly leaves 60
homeostasis 78–83
homologous chromosomes
  34, 35, 42
hormonal control
  in animals 68
  in plants 68
hormones 23, 68–72
  ADH 79
  adrenaline 81
  collagen 22, 28
  glucagon 80
  insulin 23, 80
humus 88
hunting 55
hybrid vigour 43, 46
hybridisation 46
hydrophytes 53
hypertonic organisms 50
hypertonic solutions 9
hypothalamus 79, 82, 83
hypotonic organisms 50
hypotonic solutions 9

IAA 70–1
immobilised enzyme
  techniques 65
immunity 31
immunosuppressant drugs 31
inborn errors of metabolism 66
independent assortment 34, 35
indicator species 87
indole acetic acid (IAA) 70–1
insertion 40
insulin 23, 80
intermediate species 88
interphase 34
interspecific competition 55, 56
intestinal epithelial cells 7
intraspecific competition 55, 56
inversion 40, 41
iron 74
isolation 44
isotonic solutions 9

Jacob–Monod hypothesis 64

Kleinfelter's Syndrome 42
Krebs cycle 19

lamellae 12
lateral meristems 62, 63
lead 75
leaf abscission 70
leaf pigments 10, 11
learning, in humans 58
learning curves 58
lichen 87
ligase 49
limiting factors 14
linked genes 36–7
long-day breeders 76
long-day plants 76
lymohocytes 31
lyosome 6

macro-elements 72
magnesium 72
marsupials 45
maximum sustainable yield 86